Transitions in Tension

Communication, Environment, Science and Society Set

coordinated by
Andrea Catellani and Céline Pascual Espuny

Volume 3

Transitions in Tension

Controversies and Tensions Around Ecological Transitions

Edited by

Andrea Catellani
Grégoire Lits

iSTE

WILEY

First published 2024 in Great Britain and the United States by ISTE Ltd and John Wiley & Sons, Inc.

ISTE Ltd
27-37 St George's Road
London SW19 4EU
UK

www.iste.co.uk

John Wiley & Sons, Inc.
111 River Street
Hoboken, NJ 07030
USA

www.wiley.com

Library of Congress Control Number: 2024941766

British Library Cataloguing-in-Publication Data
A CIP record for this book is available from the British Library
ISBN 978-1-78630-910-5

Contents

Author Presentation

François Allard-Huver

François Allard-Huver is an Associate Professor in Strategic and Digital Communication at the University of Lorraine, Nancy, as well as co-head of the "Praxis" research team within the Center for Research on Mediations (CREM). His work deals with the issue of environmental and health controversies around food and pesticides. He is particularly interested in the question of "affairs" in the media and public sphere. He also works on risks, sustainable development and organizations, particularly on the CSR communication and public relations strategies of various actors in the public sphere (institutions, civil society, lobbies, industry). He regularly participates in symposia and conferences on the relationship between sustainable development, risk, controversies, food and communication. A former board member of the International Communication Association (ICA), he is secretary of the "Philosophy, Theory and Critique" division, a member of the management committee of the "Communication, Environment, Science and Society" (CESS) study and research group of the French Society of Information and Communication Sciences (SFSIC) and scientific director of the Academy of Controversies and Sensitive Communication (ACCS).

Pauline Amiel

Pauline Amiel is the head of EJCAM and responsible for that school's master's program in journalism. She is also a Lecturer in Information and Communication Sciences, IMSIC, Aix-Marseille University – University of Toulon. Her work focuses particularly on solutions journalism, the local press and the roles of journalists in public spaces.

Cyrille Bodin

Cyrille Bodin holds a PhD in Information and Communication Sciences and is an associate researcher at the Research Group on Communication Issues (GRESEC), Grenoble Alpes University. His work focuses on the mediation and mediatization of science, on environmental and/or socio-scientific controversies, as well as on engagements of researchers and scientific institutions in public communication.

Andrea Catellani

Andrea Catellani is a Professor of Communication at the Catholic University of Louvain (UCLouvain, Belgium). He co-leads the "Communication, Environment, Science and Society" study and research group of the French Society of Information and Communication Sciences (SFSIC); he coordinates the international research project "Overcoming Obstacles and Disincentives to Climate Change Mitigation" (2020–2024). He has published various articles and works, notably on environmental communication and rhetoric, discourses on the social responsibility of organizations, the semiotic approach to organizations, ethics in communication, and the relationship between religion and communication in the digital world.

Mikaël Chambru

Mikaël Chambru is a Lecturer in Social Sciences at Grenoble Alpes University and scientific co-coordinator of Labex ITTEM (Innovations and Territorial Transitions in the Mountains). His research focuses on publicizing science, public controversies, and socio-environmental mobilizations.

Philippe Chavot

Philippe Chavot is a Lecturer in Information and Communication Sciences at the University of Strasbourg. He is a member of the Interuniversity Laboratory in Education and Communication Sciences (LISEC-UR 2310) and coordinates the work of the social sciences group of the Interdisciplinary Thematic Institute "Geosciences for the Energy Transition" (ITI GéoT). His work focuses on science in the media, mediation and consultation mechanisms, and socio-technological controversies.

Cécilia Claeys

Cécilia Claeys is a Professor of Sociology at the University of Perpignan Via Domitia. Her research focuses on land use planning conflicts, recreational uses of nature and environmental risks.

Amélie Coulbaut-Lazzarini

Amélie Coulbaut-Lazzarini is an Associate Professor in Information and Communication Sciences at GRESEC, Grenoble Alpes University, as well as a member of the management committee GER COMENSS. Situated in environmental communication, her current research explores the notion of human/non-human reliance and links, particularly in mid-mountain areas and in connection with the public communications of regional natural parks. Her research deals with the socio-ecological transition issues of territories and integrates the analysis of innovative and sustainable communication, awareness-raising, mediation practices and processes.

Quentin Daveau

Quentin Daveau holds a master's degree in Scientific and Technical Communication and Culture from Grenoble Alpes University.

Orélie Desfriches-Doria

Orélie Desfriches-Doria is a Lecturer in Information and Communication Sciences at Paris 8 University and a researcher at the Paragraph laboratory. Her research focuses on argumentation modeling and controversy mapping, information control and critical thinking. She is a member of the study and research group on ethics and digital technology in information-communication (GENIC).

Jules Dilé-Toustou

Jules Dilé-Toustou holds a PhD in Information and Communication Sciences at LERASS, Toulouse. He is mainly interested in analyzing discourse on web platforms, particularly from activist perspectives. He also teaches at the IUT of Castres where he teaches introductory courses in humanities and social sciences and the socioeconomics of the Internet.

Jean-Claude Domenget

Jean-Claude Domenget is a Senior Lecturer in Information and Communication Sciences at the University of Franche-Comté. He is responsible for the Conception, Creation, Mediations (CCM) division of the ELLIADD laboratory, vice-president of the professional relations commission and in charge of relations with the study and research groups (GER) within the SFSIC. His research focuses on the uses of socio-digital media, innovation through uses, particularly in the field of cycling mobility, digital professional identities and professionalization in communication professions, temporalities in information-communication. He is also developing advanced thinking regarding research ethics in a digital context within the GER on ethics and digital technology in information-communication (GENIC), which he co-founded and scientifically co-leads.

Catherine Dominguès

Catherine Dominguès is a Senior Researcher at LASTIG (ENSG/IGN and Gustave Eiffel University), a specialist in natural language processing. She has notably worked on the analysis of feelings in the narratives of Spanish Republicans and their cartographic representation. She was also part of the PARVIS project, which aimed to analyze representations of the city of the future in science fiction narratives. Her current work is based on the *Cahiers citoyens*, a corpus of volunteered citizen's words collected as part of the French *Grand débat national* and during the "gilets jaunes" movement.

Antoine Gaboriau

Antoine Gaboriau is a PhD student at CESPRA (EHESS), working on the consequences of implementing digital participation tools in local political-administrative systems in Barcelona, Geneva and Toulouse. He has also worked at Open Source Politics, a company specializing in digital participation platforms and the processing of data derived from them.

Bruno Lefèvre

Bruno Lefèvre holds a PhD in Information and Communication Sciences, attached to LabSIC-USPN. His research in socioeconomics and political economy focuses on the interdependencies between industrial strategies, public decision-making modalities, territories and civil society, focusing on the discourses and representations of these different actors.

Grégoire Lits

Grégoire Lits is a Lecturer in Media Sociology at the School of Communication and Journalism of Louvain and co-director of the Observatory for Research on Media and Journalism (ORM) of UCLouvain. His research focuses on the flow of information in public spaces in times of crisis, as well as how different public issues evolve over time. In this context, he has been conducting research for 15 years, analyzing the trajectory of various public problems such as: radioactive waste management, European policies to combat poverty, management of the SARS-CoV-2 epidemic or efforts to combat misinformation at the European level.

Catherine Loneux

Catherine Loneux is a Professor in Information and Communication Sciences at Rennes 2 University. She is a researcher at PREFics, EA 7469 (multilingualisms, representations, French-language expressions – Information, communication, sociolinguistics). The communication of organizations and the communicational dimension of the dynamics of normative construction form the structure of her research journey. Her work focuses more specifically on "corporate social responsibility", and seeks to analyze the socioeconomic, socio-legal, socio-discursive and socio-cognitive issues that these CSR phenomena underlie. Information–justification–ethical practices are studied using the analytical framework of the communication approach of organizations. To shed light on objects ranging from professional managerial environments in companies, to public institutions producing environmental and social discourses, the dynamics of standardization, the recompositions of power relations, meanings, writings and values are analyzed.

Anne Masseran

Anne Masseran is a Lecturer in Information and Communication Sciences at the University of Strasbourg, responsible for the M2 master's program in international communication, ITIRI UDS.

Jeanne Pahun

Jeanne Pahun is a PhD in political science. Her thesis, defended in 2020, is entitled "How food policies change agriculture: a comparative analysis in three regions of France". It analyzes the translation, progression and scope of the controversy embodied by the politicization of our food at the level of subnational governments. After studying the rise of the bioeconomy institutional watchword in European research policies and the operationalization of national agroecological policies, she is now a postdoctoral researcher at LISIS (INRAE), studying the role of downstream actors in the greening of agricultural producers' practices in France.

Céline Pascual Espuny

Céline Pascual Espuny is a Professor of Information and Communication Sciences at Aix-Marseille University. She works on questions of behavioral change and on the circulation of environmental issues in public space. Her research has focused on sustainable development for more than 20 years. She has thus observed human and non-human mediations, socialization and rationalization processes at work in organizations. She is a member of the IMSIC research laboratory (Mediterranean Institute of Information and Communication Sciences, EA 7492). She co-leads the "Communication, Environment, Science and Society" study and research group.

Catherine Quiroga Cortés

Catherine Quiroga Cortés holds a PhD in Information and Communication Sciences at Toulouse III – Paul Sabatier University, attached to the Laboratory of Applied Studies and Research in Social Sciences (LERASS). In her doctoral thesis, she examines local information, considering its production and circulation as the result of relational and communicational processes. She is interested in the role played by socio-digital networks within local information ecosystems and questions the impact of these digital devices on the restructuring of the balance of power underlying the deployment of said processes.

Erika Riberi

Erika Riberi is a Lecturer in Information and Communication Sciences at the Faculty of Science of Aix-Marseille University, and a researcher at IMSIC (Aix-Marseille University – University of Toulon). She is interested in the presence and circulation of scientific discourses in public space (environmental communication, scientific mediation, science–society relations). She is responsible for the first year of the master's course in scientific information and environmental mediation.

Brigitte Sebbah

Brigitte Sebbah is a Professor of Information and Communication Sciences at Toulouse III – Paul Sabatier University, within LERASS. Her research focuses on digital journalism, changes in journalistic practices, online information and social movements.

Yeny Serrano

Yeny Serrano is a Lecturer in Information and Communication Sciences at the University of Strasbourg and attached to LISEC (UR 2310). She is interested in the information discourses of the mass media and their place in contexts of controversy or armed conflicts.

Natacha Souillard

Natacha Souillard is a Lecturer in Information and Communication Sciences at Aix-Marseille University, within the IMSIC laboratory. Her research focuses on citizen participation and online social movements, the circulation of representations, discourses and narratives of environmental and ecological issues.

Bi Mathieu Tra

Bi Mathieu Tra holds a PhD in Information and Communication Sciences at the University of Franche-Comté and is a member of the ELLIADD laboratory. His thesis is titled "Polyphony on socio-digital media: the case of interactions on YouTube about the Amazon forest fires". It was supervised by Jean-Claude Domenget and Orélie Desfriches-Doria.

Kimberley Vandenhole

Kimberley Vandenhole is a PhD student at the Socio-Environmental Dynamics Research Group (SONYA) at the Free University of Brussels and the Research Center for Sustainable Development at Ghent University. She studies environmental phenomena from socio-political and discursive perspectives in the context of morality and societal transformations. Interested in the socio-political and philosophical dimensions of environmental issues, she is currently focusing on analyzing the discourse of eco-shaming.

Albin Wagener

Albin Wagener is a Teacher–Researcher in language sciences. Specializing in digital, media and climate discourse analysis, he is attached to Rennes 2 University and INALCO. He also acts as a corpus study expert for foundations and companies.

Jean Zoungrana

Jean Zoungrana is an Associate Lecturer at the Faculty of Social Sciences at the University of Strasbourg. He is attached to the SAGE (Societies, Actors, Government in Europe) laboratory (UMR 7363) and a member of the social sciences group of the Interdisciplinary Thematic Institute "Geosciences for the Energy Transition". His work focuses on governmentality and citizen action in the context of consultation and controversy.

Introduction

Controversies and Tensions Around Ecological Transitions: For an Infocommunicational Approach

In recent years, the metaphor of the "transition", which suggests a movement, a priori soft and smooth, from one state to another, has taken a growing place in both media and scientific discourses on the evolution of human societies. Beyond the apparent peacefulness associated with it, this term often appears in connection with issues associated with quite lively tensions and controversies, whether related to the ecology and habitability of the planet, energy policy, even the evolution of our social protection models or the evolution of our communication tools (when we talk about a society's digital transition). It is this apparent contradiction between an imaginary of gentle evolution, of a natural change of state and the strong tensions that these changes arouse in the social body that this book attempts to elucidate by presenting different works of research from the information and communication fields, and more broadly the humanities and social sciences, focusing on the environmental and energy sphere.

The *Dictionnaire critique de l'anthropocène* (2020, p. 780) takes from Chabot (2015) the idea that we have entered the "age of transitions", defined as a "fundamental reconfiguration of the functioning and organization of the system, faced with a tipping point". But the notion of an ecological transition, unlike others such as "demographic transition", indicates not so much a phenomenon as an "intention", a "watchword prescribing practices" (ibid.), used by a wide range of actors (activists, institutional, professional, technical, political, scientific), just as for the "digital transition".

Introduction written by Andrea CATELLANI and Grégoire LITS.

Like the latter, the ecological transition describes a process of "transformation during which a system passes from one regime of equilibrium to another" (Bourg and Papaux 2015, p. 780, cited by Monnoyer-Smith (2017)). The term is particularly linked to the "transition towns" movement and the transition network, a movement initiated by Rob Hopkins in England in 2005. More broadly, the notion of transition has become one of these "formulas" (Planque 2010) used today in discourse dealing with the ecological situation, its consequences and perspectives for action in the political, media, economic, social, spiritual and cultural spheres (as well as the scientific, with recent research on sustainability transitions). This formula indicates a complex set of values, scientific notions, projects, projections, actions and practices involved with the transition from the current state towards a more "sustainable", lasting and desirable condition for humans and living beings, thus partly replacing the semantics of sustainable development (Theys 2020), and also implying "the resorption of social inequalities or environmental injustices" (*Dictionnaire critique de l'anthropocène*, 2020, p. 780), as recalled by the notion of "just transition" also present in the text of the Paris climate agreement. In France and Spain, for example, there exist Ministries of Ecological Transition, even if the names quickly change.

The positive, dynamic and promising connotations of this type of expression (different from those of other expressions such as collapse, therefore with potential for euphemization) certainly help the dissemination and use of these formulas. Different actors therefore appropriate this expression within different logics and perspectives, such as those of long-term public policies, that of associations and movements oriented towards the commons and the local, or even that of large companies (and more particularly in the energy sector). Transition makes it possible to build a positive framework, linked to the discourse of innovation, of design thinking, of participation, etc. This diversity and heterogeneity – which does not deny the existence of a "family resemblance" between the different uses – is interrogated by research in information and communication or even discourse analysis.

For its part, the expression "energy transition" is defined by the World Energy Council (2014) as "a significant structural change in an energy system". The current transition to renewable energy is the latest energy transition, and it is driven directly by the need to reduce greenhouse gas emissions in this era of ongoing global climate and environmental crisis. However, as the *Dictionnaire critique de l'anthropocène* reminds us, "the energy transition is not determined in its solutions, nor in the modalities of its implementation" (2020, p. 783). Different, more or less radical models and perspectives intersect, integrating ecological issues to a greater or lesser extent. From this point of view, the energy transition and its challenges "question our relationship to the Earth" (*Dictionnaire critique de l'anthropocène* 2020, p. 784) and therefore the self-definition of societies. Here too, there is material for work in information and communication sciences (Gilbert et al. 2019) and more widely in the humanities and social sciences.

This book, the result of a colloquium held at the Catholic University of Louvain (UCLouvain), Belgium, on December 16 and 17, 2021, aims to question the discourses and communicational phenomena related to ecological and energy transitions – we explicitly use the plural to indicate the complexity of the subject in question. The aim is to focus on the polemical dimension of the communication surrounding these transitions – controversies, polemics, discussions, debates – which mobilizes verbal language and any other type of semiotic mechanism (particularly images, both still and moving). We thus place ourselves within the lineage of works which have successfully defined and mobilized the notion of controversy to analyze the processes of social evolution, such as those of Allan Mazur on the dynamics of technological controversies (Mazur 1981), those which have developed from French pragmatic sociology (Boltanski and Thévenot 1991; Callon et al. 2001; Latour 2006), or even more recent ones which have developed in the field of environmental information and communication in the French-speaking world (see, for example, Carlino and Stein (2019)).

Transitions and the tensions and controversies surrounding them are objects of multimodal discursive appropriation (textual, verbal, visual) by multiple actors (Audet 2014). Information and communication sciences are then called to contribute to better understanding of the semiotic, rhetorical, narrative and discursive forms mobilized in these discourses. The controversies around transitions are indeed powerful "semiotic engines", which activate the production and circulation of discourses. The notion of controversy is first of all a "descriptor" allowing the consideration of a set of particular real social situations, which embody "transitions" based on the communication activities of social actors. In this sense, controversies are above all social phenomena, interactional episodes. As Pierre Lascoumes notes (Lascoumes 2010, p. 172), "[they] can be defined as sequences of discussion and confrontation between divergent points of view on a subject". Controversy is a moment in history during which actors create confrontations between their points of view on a social issue.

The definition of controversy as a process of construction suggests that it fulfills another objective. Beyond its descriptive character, it is also an analytical operator bearing a relationship to particular expert knowledge. It allows the researcher to analyze society, to construct an object of study from a particular angle and following a defined method. This method is often associated with that of "deploying controversies" (Latour 2006, p. 33). This methodological posture rests on a basic foundation close to the "actor-network" posture proposed by Latour, which "claims to be better able to find order after having let the actors deploy the whole range of controversies in which they find themselves immersed. [...] In other words, the task of defining and ordering the social must be left to the actors themselves, instead of being taken over by the investigator" (Latour 2006, p. 33). In this sense, it is a question of "unfolding" certain particular controversies in order to open certain "black boxes" linked to expert or lay discourses on "transitions". It is a question of how experts and laypersons interact and enter into dialogue when it comes to defining the issues of transitions.

Another objective of the notion of controversy is more normative and assumes the positive nature for society of the deployment (by actors and researchers) of controversies in different fields of public action (Callon et al. 2011). In this sense, controversies are seen as a democratic procedural model characterized by a double questioning of the delegation of decisions to scientists and politicians. As a collective decision-making procedure, they would then be "paths to a better articulation between science and society" (Lascoumes 2002, p. 72). Some research highlights how controversies linked to transition issues, as a decision-making procedure, come to produce solutions that are considered socially and technically "robust" (e.g. Anderies et al. 2004) or with "social acceptability" (e.g. Wüstenhagen et al. 2007). They also question the links between the evolution of the debate in the public and media sphere, and the evolution of the modes of constitution of democratic legitimacy (Rosanvallon 2008). Alongside classic approaches centered on the analysis of public participation in democratic or sociotechnical decisions, others question these notions of robustness and social acceptability based on the study of communication processes or participatory or deliberative processes underlying controversies linked to transition issues.

Finally, it should be noted that environmental controversies, which are at the heart of debates related to "transitions", have particular characteristics: the long time necessary for their resolution (Lascoume 1994; Blanck 2016), their transversal and intersectoral character (Lascoumes 1994) and finally the change in decision-making scale that they require as well as their multi-stakeholder character (Gouldson 2009; see also Carlino (2018) and Lits (2020)). These characteristics make them an important field for democratic innovation and for the development of an important set of "new instruments of public action", as well as new modes of production of legitimacy. It is therefore interesting to look at the communicational aspects of this management of time, of this intersectoral, multi-actor and multi-level as well as the transnational character of environmental controversies.

The chapters comprising this volume develop some of the axes and perspectives mentioned above. The first two immerse us in the French agricultural sector, a world affected by tensions linked to the demands of the ecological transition.

In Chapter 1, François Allard-Huver questions the mediatization and circulation of the concept of *agribashing* and the semiopolitical implications of its use by certain actors in the French agricultural world, in a context of ecological and agricultural transition. More precisely, the author questions, firstly, the way in which the concept of agribashing emerged and established itself as a real formula (Krieg-Planque 2010; Simon 2015) in defending a certain agricultural model. He then observes the media coverage and circulation of this term, to question, finally, how it appears to disqualify those actors in civil society who mobilize a discourse critical of intensive agriculture and promoting the ecological transition. The corpus is based on the study of the French daily and regional press between 2018 and 2021, as well as on semiopolitical analysis of the supporting discourses of agribashing and the material forms of their inscription in the public sphere: reports, official documents and communication materials produced as part of the establishment of the "Déméter" cell, created within the French national gendarmerie to prosecute attacks and intrusions against agricultural operations. Does talk of agribashing amount to placing discourses around changes in the agricultural model and the agricultural transition in a comminatory and agonistic register? Do the polemics over critical discourses and civil society actions against part of the agricultural world reflect opposing visions of the ecological transition that nothing seems able to reconcile?

Chapter 2, written by Jeanne Pahun, takes up the theme of agribashing. To fight against this supposed collective disavowal of the agricultural world, hegemonic representatives of the agricultural sector, such as the FNSEA federation, use different strategies. One of them is to participate in the currently booming construction of urban food policies in order to convey a positive image of agriculture to urban centers, places where the population is concentrated and opinion is formed. Even if the type of agriculture supported within urban food policies (organic and diversified farming, run by new rural residents, including farmstays, etc.) diverges from the agricultural development promoted by the FNSEA and its representatives, the latter have a strategic interest in participating and collaborating in it. For them, it is a question of working on constructing a new image of the agricultural world, closer to the new expectations of citizens. This communication strategy contributes to the establishment of "showcase agriculture" around the cities displaying a transformation of the agricultural sector, which remains marginal at the level of the national territory. This chapter is based on a political science thesis on the construction and deployment of agricultural components of territorial food policies, particularly those of the cities and metropolises of Rennes, Montpellier and Strasbourg. Around 30 interviews conducted between 2017 and 2019 with key players in these policies constitute the empirical material for this work.

The following three chapters provide a better understanding of the tensions that exist when the ecological transition takes place in specific territories, where different interests and values intersect and sometimes clash, particularly around development projects.

In Chapter 3, through the analysis of 15 contested development projects in Belgium and France, Bruno Lefèvre shows that the territorialization strategies of industrial players in sectors as varied as the digital or leisure industries, supported by public decision-makers, lead to their own questioning by civil society. Phenomena of repoliticization of local territories and their actors situate these projects, initially presented as relatively anecdotal, within a complex set of contemporary social and environmental challenges. The meaning of these projects with regard to these major issues and integrated at scales other than the local territory becomes subject to ideological conflicts which the consultation bodies fail to animate. Discourses, media and digital social networks contribute to making these conflicts visible, to identifying residents and activists, and to enabling forms of cooperation on a national and international scale, in favor of a new balance of power between civil society and public decision-makers.

Actors in the French mid-mountain areas, through their environmental communication highlighting their role as mediators, increasingly position themselves as leaders in a transition focused on conciliations and links. Chapter 4, written by Amélie Coulbaut-Lazzarini, aims to understand how such communication can enable moving from a perception of environmental crisis, of tension, to the construction of a harmonious future, between individuals, groups of individuals and linked non-human actors. From a methodological point of view, the results presented in this chapter are based, on the one hand, on the analysis of the communication documents produced or co-produced by the Bauges RNP (Regional Natural Park), and, on the other, by field surveys. The programs studied, supported by RNPs, place the mid-mountain environment at the heart of mechanisms of environmental communication. Human/non-human interactions are involved in the very structuring of communication mechanisms, where the ability to perceive otherness is the basis of the envisaged social project. The communicational strategies enacted by actors, particularly public actors, on issues of the relationship to the living world are based on explorations of types of mechanism that create links between the public and the natural environment. These focus not on tension, but on living together, on cooperation. They anchor themselves in experience, in the sense of the philosopher John Dewey, in order to explore links. To return to the path pursued in this chapter, it seems that the transition investigated here is a shift from a perspective of continuous growth in our Western societies to a perspective of balanced systemic co-evolution. The study of the mechanisms used by the actors mentioned situates environmental communication in a paradigmatic transition where relatedness, as a complex system of links, becomes the central element of societal dynamics.

Mikael Chambru, Cécilia Claeys and Quentin Daveau author Chapter 5, dedicated to the French mountain region. This chapter, in fact, analyzes the trajectory of a scientific tourism project as a possible lever for territorial transition, responding to the double observation of the exhaustion of the "white gold" model and the mountain region's early display of signs of climate change. This involves the creation of a museographic space within the Dévoluy massif, in the Hautes-Alpes department. Based on a qualitative field survey conducted in 2019 and 2020, this chapter shows how a potentially eco-compatible form of territorial innovation based on revitalization of the historic village through the enhancement of a local cultural, environmental and scientific heritage is sliding towards a strategy of strengthening modernist tourism centered on the alpine ski resort.

Chapter 6, written by Catherine Quiroga Cortés, looks at a case of digital mobilization, in the context of a local energy controversy, in which the content of the more "traditional" media plays an important role. Indeed, rejecting the framing produced by the regional media, opponents of the offshore wind farm in Saint-Brieuc Bay (France) are turning to Facebook to publicize their own account of the controversy. However, within the various discussion spaces found in the social network, content from the said media is circulated. In order to understand the place occupied by regional media discourse in construction of an oppositional framing, the author conducts a comparative study of the two discourses based on a textometric analysis (using the IRaMuTeQ software). She notes that, despite a significant difference in the way that environmental impacts of the park are reported, and an activity resembling the gatekeeping performed by Facebook users, the regional media discourse seems central in promoting the words and actions of protest.

The following two chapters continue to explore the mobilization of different digital media in the context of controversies around ecological transitions.

Chapter 7, written by Erika Riberi and Pauline Amiel, focuses on the democratic role of the news media, which is one of the predominant actors in the public space in which controversies over ecological transitions take place. He explores more specifically the treatment of transitions through a booming media object – the podcast, and a particular approach: solutions journalism. For this purpose, a corpus of 10 native podcasts, broadcast between 2017 and December 2021, was compiled and analyzed. The updated results firstly confirm a strong trend towards depoliticization and individualization of these issues, in accordance with the analyses of Comby (2019). Above all, they make it possible to uncover a third dimension that is particularly fruitful for addressing possible courses of action for working towards environmental transitions: the economic dimension. Four types of action combining to a greater or lesser degree the individual, political and economic dimensions are thus distinguished in the corpus: 1) the so-called "inspiration" course of action, which relies only on individualization; 2) that which combines the individualizing and the economic, advocating the individual creation of products or businesses promoting transitions; 3) the course of action combining individualization with a more or less strong political coloring; and finally, 4) a course of action that covers all dimensions and synthesizes all the various courses of action by presenting the enterprise as a new place of general interest.

The purpose of Chapter 8, written by Bi Mathieu Tra, Jean-Claude Domenget and Orélie Desfriches-Doria, is to shed light on some of the obstacles and methodological issues encountered in the study of an ecological controversy on YouTube, that of forest fires in the Amazon. It is therefore a chapter focused on the methodological dimension: by combining exploration and representation using graphing tools, a socio-informatics approach to controversies (Francis Chateauraynaud) and screen writing (Emmanuel Souchier), the authors propose a solution for the qualitative study of controversies on YouTube. Based on this methodology, a *crawl* of hypertext links using Hyphe software, for the 2019 period, allowed the authors to establish a cartography and categorization of the types of actors engaged on the Web in exchanges on Amazon forest fires. This operation also showed the key role of technical mediation in the co-construction of interactions, while the socio-informatics approach allowed attributes to be given to actors, to the key events and identification of a discursive polarization opposing different opinions concerning the Amazon forest fires. Starting from a structural approach to relationships, the authors arrive at a qualitative analysis of the forces involved.

The following three chapters show the results of textual and discourse analysis approaches applied to issues related to controversies over ecological transitions.

The objective of Chapter 9, by Kimberley Vandenhole, is to explain, with empirical evidence, the specific interpretation of the notion of "sustainable transition" advocated by a particular environmental discourse, that of eco-shaming. This interpretation is reconstructed from a discourse analysis of 2,155 documents distributed in Belgium between 2018 and 2021. The documents analyzed are newspaper articles, written parliamentary questions published by the federal, Brussels, Flemish and Walloon parliaments, and newsletters from environmental movements. Every speech contains storylines dealing with three frameworks: diagnostic, prognostic and motivational frameworks. These frameworks correspond to the three phases that characterize a transition: a given state of the system, a process of change and a new state of the system. Thus, by analyzing the storylines of the discourse of eco-shaming, the specific interpretation of "sustainable transition" advanced by this discourse is identified and described. The chapter ends with a comparison between the notion of "sustainable transition", as advocated by the discourse of eco-shaming and that advocated by other environmental discourses.

Albin Wagener, author of Chapter 10, focuses his attention on discourse about the electric car. At the heart of sensitive climate issues, transport is particularly interesting, insofar as it materializes economic, social and cultural tensions. The electric car is no exception to this rule and has received specific media attention, due to the humanitarian and ecological controversy surrounding the production and recycling of batteries. This chapter aims to study the representations that circulate in the socio-digital environment, through a corpus study based on textual data collected from social networks such as Facebook, YouTube or Twitter, using tools that aid in textual analysis.

Chapter 11, written by Brigitte Sebbah and Natacha Souillard, proposes a contribution to the understanding of the narrative of environmental issues, by "unfolding" what the presence on the media agenda of the notion of "ecological transition" leads to in terms of framing, and by showing the extent to which analysis of these frameworks sheds light on the processes of reformulation or requalification of related issues, in a context of territorialization of public policies. To investigate these questions, the authors conducted a contrastive study, using the IRaMuTeQ software, of the dominant media framings of the notion of ecological transition, in two types of mainstream press, the French national (NDP) and regional (RDP) daily press. In three stages, this study firstly questions the stability and circulation of the notion, then what this notion covers in terms of framing and meaning, before questioning the articulation of this perimeter of meaning with the dimension of "doing" public and political action. The results of this analysis trace a notional perimeter sensitive to the social and political situation, but dominated by economic and managerial registers, between episodic and thematic treatment, with phenomena of prioritization. In addition, the media treatment oscillates between two approaches to ecological issues, that actualize, in the narrative of ecological transition, a phenomenon of competition rather than real convergence between the paradigms of weak sustainability and strong sustainability.

The following two chapters emphasize the involvement of different types of organizations in the transition and the resulting discursive productions.

Chapter 12, by Catherine Loneux and Céline Pascual Espuny, seeks to understand, through the prism of communication by public policy actors, ecological actors, companies and citizens, the articulation between regulations, decision-making and organizational changes. Since the spring of 2021, environmental issues have come increasingly to the foreground of the European and French institutional agendas. Marked by a project of green growth, ecological transition and renewable energy, affecting more and more fields (automotive, food, agriculture, technologies, etc.), the French climate action plan was launched, based on a "green budget" and complemented by an investment program for the future, and a plan for the ecological transition of VSEs and SMEs. These evolutions and recompositions offer a shift or at least a societal reconfiguration towards more ecology, as well as new norms, and result in an accompanying discourse that has attracted the authors' attention. The chapter focuses especially on labels, formulated and communicated within the framework of this state activism, analyzed as an expression of "normativity" linked to the transition. These labels are observed by mobilizing "organizational communication approaches" (OCAs).

Chapter 13, written by Jules Dilé-Toustou, synthesizes his thesis work in information and communication sciences on the investment in the Web of ecological organizations. It focuses on the channels of distribution of French ecologist discourses, namely the dominant websites and social-digital networks (SDN) (mainly Facebook, Twitter and Instagram). Using a diversity of organizations representative of the French ecologist milieu, it makes it possible to distinguish which organizations and, indeed, what types of discourse are disseminated on different web platforms. We thus learn that the most radical discourse is disseminated very little on the SDNs, but persists via websites. Furthermore, the most moderate environmentalist speeches are widely published on the dominant SDNs, thanks to the investment in these platforms of contemporary environmentalist organizations.

The last two chapters allow us to observe closely, from a discursive and communicational point of view, the issues of citizen participation and involvement in the face of ecological challenges.

The research presented by Antoine Gaboriau and Catherine Dominguès in Chapter 14 analyzes the situations of enunciation in two corpora of digital citizen contributions produced in 2020: the proposals addressed to the French Climate Citizen Convention and the results of the digital consultation organized by a group of French parliamentarians, The Day After. This research work is situated in the field of discourse analysis, defined as analysis of the relationship between text (contributions) and context (the discursive community created by the device). It seeks to interpret the regularities of language observed in contributions to the two digital citizen consultations, and more specifically the management of enunciation through analysis of the personal pronoun *we* and the possessive determinant *our*. This contrastive analysis then makes it possible to precisely identify the positions and purposes specific to these two discourses. This ultimately leads into an effort to politically interpret the two corpora.

Chapter 15 takes us to Alsace, where the rapid development of deep geothermal (DG) techniques has led to a socio-technical controversy concerning the presence of power plants, or future power plants, in territories with a high population density. A form of non-intermittent renewable energy, DG also presents certain risks, particularly of a seismic nature. As part of this chapter, the authors (Cyrille Bodin, Philippe Chavot, Anne Masseran, Yeny Serrano and Jean Zoungrana) present the results of a series of group interviews, of the focus group type, conducted with local citizens concerned by these projects. Thus, the controversy raised by some residents, local elected representatives or associations primarily involves, more than the technical modalities of the projects, the largely counter-productive modalities of implementation of public policies for energy transition: a popularization policy directly impedes the possibility of citizen participation. The authors thus conduct a critical reflection on the need for a communicational and epistemological transition to accompany environmental policies, to make them effective and socially responsible.

The chapters, here briefly introduced and then left for the reader to explore, show – among other things, we hope – the importance of in-depth consideration of the communicational and discursive aspects of the energy and ecological transitions. This volume, in the diversity of its approaches and objects, shows in particular the usefulness of situated studies, which seek to account for the complexity of specific and local situations by showing how transitions are embodied in social worlds, and which may usefully complement more general and global work. In-depth analysis, mobilizing the best tools of the humanities and social sciences, may be a useful contribution, we hope, to accelerating the changes needed to manage the challenges of our time.

I.1. The "communication, environment, science and society" research group

This volume is an expression of the activities of the "communication, environment, science and society" study and research group, which is part of the French Society of Information and Communication Sciences (SFSIC). The group is a space for encounter and research in environmental communication, defined by the International Environmental Communication Association (IECA) as "all of the diverse forms of interpersonal, group, public, organizational, and mediated communication that make up the social debate about environmental issues and problems, and our relationship to the rest of nature".

This group aims to offer an open space for dialogue between scientists, society and professional communities in order to increase the influence of research conducted around increasingly important themes in our society. All the researchers forming this group wish to build skills and knowledge useful for the "ecological transition" which has become essential in the era of the Anthropocene. The thematic scope of the group includes the different communicational aspects of notions related to environmental concerns: ecology, sustainable development, ecological transition, risk, scientific discourse and popularization, scientific or museum communication, mobilization, crises, the discourses and communicational activities of public and private actors, both commercial and non-commercial, at the local, national and international levels, using any type of technology, support and device. All approaches and methodologies for research in information and communication sciences may be used.

The group regularly organizes scientific activities, in particular annual colloquia. Its activities can be monitored via the "research notebook" on the Hypotheses platform[1]. The group also has a collection on the HAL[2] scientific platform and a public group on Facebook.

This volume is part of an initiative of promotion of research carried out within the group's scientific scope and more broadly on environmental themes. Indeed, as already mentioned, the chapters comprising this book are a selection of the interventions presented during the 2021 annual scientific colloquium of the study and research group, entitled "Transitions in tension – controversies and tensions around ecological transitions", which took place in Louvain-la-Neuve in Belgium (see the website of the colloquium)[3]. Each chapter was subject to peer review before being included in this volume: the latter therefore constitutes an original work in its own right, and it aims to advance knowledge on communication issues and social aspects of ecological and energy transitions.

References

Anderies, J.M., Janssen, M.A., and Ostrom, E. (2004). A framework to analyze the robustness of social-ecological systems from an institutional perspective. *Ecology and Society*, 9(1) [Online]. Available at: www.jstor.org/stable/26267655 [Accessed 1 December 2020].

Audet, R. (2014). The double hermeneutic of sustainability transitions. *Environmental Innovation and Societal Transitions*, 11, 46–49. doi: 10.1016/j.eist.2014.02.001.

Bernard, F. (2018). Les SIC et l'"anthropocène" : une rencontre épistémique contre nature ? *Les Cahiers du numérique. Pluralisme épistémologique et conceptuel en information-communication*, 14, 31–66.

Blanck, J. (2016). Gouverner par le temps. *Gouvernement et action publique*, 5(1), 91–116.

Boltanski, L. and Thévenot, L. (1991). *De la justification. Les économies de la grandeur*. Gallimard, Paris.

Callon, M., Lascoumes, P., Barthe, Y. (2001). *Agir dans un monde incertain – Essai sur la démocratie technique*. Le Seuil, Paris.

Carlino, V. (2018). Temporalités de la controverse sur le nucléaire en Lorraine. Temps "long" des désaccords, temps "infini" des déchets radioactifs. *Questions de communication*, 34(2), 155–172.

1 See: https://comenvironnement.hypotheses.org/.

2 See: https://hal.science/GR-CESS.

3 See: https://transition2021.sciencesconf.org/.

Carlino, V. and Stein, M. (eds) (2019). *Les paroles militantes dans les controverses environnementales. Constructions, légitimations, limites.* PUN-ÉDUL, Nancy.

Catellani, A. (2009). La communication environnementale interne d'entreprise aujourd'hui : dissémination d'un nouveau "grand récit". *Communication et organisation*, 36, 178–188.

Catellani, A., Pascual Espuny, C., Malibabo, P., Jalenques Vigouroux, B.J. (2019). Les recherches en communication environnementale. État des lieux. *Communication*, 36(2) [Online]. Available at: http://journals.openedition.org.

Catellani, A., Pascual Espuny, C., Malibabo Lavu, P. (2022). Environmental communication research in the French-speaking world. In *The Handbook of International Trends in Environmental Communication*, Thakahashi, B., Metag, J., Thaker, J., Comfort, S.E. (eds). Routledge, London, 268–284.

Chabot, P. (2015). *L'âge des transitions*. Presses Universitaires de France, Paris.

Collectif (2020). *Dictionnaire critique de l'anthropocène*. CNRS Éditions, Paris.

Comby, J.-B. (2019). Retour sur la dépolitisation des enjeux écologiques. In *Manuel indocile de sciences sociales. Pour des savoirs résistants,* Fondation Copernic (ed.). La Découverte, Paris, 470–480.

Galaup, X. (2012). Introduction – Mode d'emploi. In *Développer la médiation documentaire numérique*, Galaup, X. (ed.). Presses de l'ENSSIB, Villeurbanne.

Garbiès, C. and Fabre, I. (2012). Définition et enjeux de la médiation numérique documentaire. In *Développer la médiation documentaire numérique*, Galaup, X. (ed.). Presses de l'ENSSIB, Lyon.

GIEC (2001). Changements climatiques 2001. Summary. GIEC, Geneva.

Gilbert, C., Smith, H., Bidwell, D., Smythe, T., Moore, A., McCann, J., Miller, E. (2019). Gatekeeping and communities in energy transition: A study of the Block Island Wind Farm. *Environmental Communication*, 13(8), 1041–1052. doi: 10.1080/17524032.2018.1561484.

Gouldson, A. (2009). Advances in Environmental Policy and Governance. *Environmental Policy and Governance*, 19(1), 1–2.

Krieg-Planque, A. (2010). La formule "développement durable": un opérateur de neutralisation de la conflictualité. *Langage et société*, 4, 5–29.

Lascoumes, P. (1994). *L'éco-pouvoir. Environnements et politiques*. La Découverte, Paris.

Lascoumes, P. (2002). De l'utilité des controverses socio-techniques. *Journal international de bioéthique*, 13, 68. doi: 10.3917/jib.132.0068.

Lascoumes, P. (2010). Controverse. In *Dictionnaire des politiques publiques*, Boussaguet, L., Jacquot, S., Ravinet, P. (eds). Presses de Sciences Po, Paris, 172–179 [Online]. Available at http://www.cairn.info/article.php?ID_ARTICLE=SCPO_BOUSS_2010_01_0172 [Accessed 16 July 2012].

Lascoumes, P. and Le Galès, P. (2006). *Sociologie de l'action publique*. Armand Colin, Malakoff.

Latour, B. (2006). *Changer de société, refaire de la sociologie*. La Découverte, Paris.

Lits, G. (2020). The European governance of radioactive waste. A story of stakeholders' involvement. In *EU Environmental Governance: Current and Future Challenges*, Orsini, A. and Kavvatha, E. (eds). Routledge, London, 79–97.

Luyckx, C. (2020). *Écophilosophie. Racines et enjeux philosophiques de la crise écologique*. Académia, London.

Mazur, A. (1981). *The Dynamics of Technical Controversy*. Communications Press, London.

Monnoyer-Smith, L. (2017). Transition numérique et transition écologique. *Annales des Mines – Responsabilité et environnement*, 3(3), 5–7. doi: 10.3917/re1.087.0005.

Rosanvallon, P. (2008). *La légitimité démocratique. Impartialité, réflexivité, proximité*. Le Seuil, Paris.

Simon, J. (2015). Formule. *Publictionnaire. Dictionnaire encyclopédique et critique des publics* [Online]. Available at: http://publictionnaire.huma-num.fr/notice/formule [Accessed 24 January 2023].

Theys, J. (2020). Le développement durable face à sa crise : un concept menacé, sous-exploité ou dépassé ? *Développement durable et territoires*, 11(2) [Online]. Available at: http://journals.openedition.org/developpementdurable/17468 [Accessed July 2020].

Wehrling, Y. (2020). Stopperons-nous la dégradation de la nature au niveau mondial ? *La biodiversité entre urgences et complexité*, 100.

Wüstenhagen, R., Wolsink, M., Bürer, M.J. (2007). Social acceptance of renewable energy innovation: An introduction to the concept. *Energy Policy*, 35(5), 2683–2691.

1

Challenges to the Agroecological Transition: Can We Talk About "Agribashing" in France?

1.1. Introduction

In October 2019, the French Ministry of the Interior announced the creation of a working group entitled "Déméter" under the gendarmerie. According to Christophe Castaner, Ministry of the Interior at the time, the "Déméter task force" is a response to "agribashing", distrust and hostility against French farmers (Ministère de l'Intérieur 2019). Created under pressure from the two main agricultural unions, the National Federation of Agricultural Holders' Unions (FNSEA) and Young Farmers (JA), the missions of this "task force" are to help farmers against theft, detect new threats, promote judicial responses as well as, more controversially, to prevent and monitor "actions of an ideological nature, whether simple symbolic actions of denigrating the agricultural environment or brutal actions with material or physical repercussions" (Gendarmerie Nationale 2019). The newly created working group is also working in coordination with 10 regional "agribashing observatories" which are to assess the nature of "the stigmatization of agricultural and livestock activities and the sale of products from this sector" (Seine-Maritime Department 2020).

Indeed, in recent years, the agricultural world has faced increasingly strong societal demands to move towards an agricultural model more respectful of the environment and in line with the challenges of ecological transition: municipal decrees in favor of buffer zones, a governmental desire to ban certain synthetic pesticides such as glyphosate or neonicotinoids, or even laws on animal welfare. Though the agricultural world does not ignore these injunctions to change their

---

Chapter written by François ALLARD-HUVER.

model, other preoccupations have come to intersect with these issues, whether economic, like the significant decline in the number of farmers and the erosion of their incomes (Chardon et al. 2020; Delame 2021) or psychosocial, such as a higher suicide rate than the general population (MSA 2020; AFP and Le Monde 2021). Considering their efforts sufficient in terms of health and the environment (Le Hesran 2020), some actors in the agricultural world seem susceptible to a growing malaise and a feeling of constant denigration, culminating with the FNSEA and JA's national campaign: "France, do you still want your farmers?" (FNSEA 2019). A paradoxical sentiment in view of the fact that most opinion studies show not only the attachment and confidence of French consumers towards "small producers" and "farmers" in general (ANIA and TNS-Sofres 2013; FrenchFood Capital 2018), but also their willingness to support them – by paying more – to adopt more rational models ensuring food independence in France (BVA and Crédit Agricole 2021; IFOP and Ouest-France 2022). Thus, in a context of ideological confrontation over models of agricultural transition, the concept of "agribashing" is controversial.

In this chapter, we question the mediatization and circulation of the concept of "agribashing" and the semiopolitical implications of its use by certain actors in the agricultural world, in a context of ecological and agricultural transition. More precisely, we question, firstly, the way in which the concept of "agribashing" emerged and established itself as a real "formula" (Krieg-Planque 2009; Simon 2015) in defense of a certain agricultural model. We then observe the media coverage and circulation of this term, to finally question how it seems to disqualify those actors in civil society who promote a discourse critical of intensive agriculture and promoting the ecological transition. Our corpus is based on the study of the French daily (n = 121) and regional (n = 1356) press between January 2018 and October 2021, as well as on semiopolitical analysis of the supporting discourses of "agribashing" and the material forms of their inscription in the public sphere: reports, official documents and communication materials produced as part of the establishment of the "Déméter task force".

Thus, in line with the questions raised in information and communication studies around the issues of environmental communication as well as around the increasing complexity of discourses relating to changes in models (Bernard 2018; Catellani et al. 2018), does talking about "agribashing" amount to framing discourses around changes in the agricultural model and the agricultural transition within a comminatory and agonistic register? Do the polemics over critical discourses and civil society actions against part of the agricultural community reflect opposing and unreconcilable visions of the ecological transition?

1.2. "Agribashing", a formula?

Mentioned in March 2016 during a debate at the *Salon international de l'agriculture* in Paris (Agriculture & Environnement 2016), by the editor of the "Agriculture & Environnement" newsletter, "agribashing" is a portmanteau made up of "agriculture" and "bashing", meaning an attack or harassment, and also implies any form of significant denigration. After circulating for a while on Twitter, it was popularized in 2018 by the "expert consultant" Eddy Fougier in his report "The agricultural world facing the challenge of 'agribashing'", where he concluded that "agribashing" was a real thing, corresponding, according to him, to an "intensification, expansion, radicalization and greater visibility of the questioning of the conventional mode of agricultural production" (Fougier 2018, 2020). The concept was then widely popularized by FNSEA and other journalists defending the conventional food and agricultural model. Analysis of the documents supporting the creation of the "Déméter" task force is thus informative about the nature of the challenges facing the agricultural world.

Indeed, behind the creation of the task force and the observatories lies the ideological postulate shared by the proponents of the agri-food and agro-industrial system and their allies, such as pesticide producers, that the French agricultural and food system is facing a rising wave of hostility, a "systematic denigration on social networks and in the media of 'conventional' farmers" (Rass 2020). This hostility is said to result in violent actions against actors in the agricultural world as well as in discourses challenging the productivist model. Among these "actions", anti-speciesist activists are particularly targeted, as well as other environmental groups in civil society carrying out "ideological actions", for example, against the use of pesticides, and therefore automatically considered "extremists".

For others, particularly environmentalist and anti-speciesist associations, the use of the term is a strategy of backlash designed to divert the attention of public opinion and government resources in favor of a single model and to muzzle opposition (Astier and Rimbert 2020; Chapelle 2020). Furthermore, for Stéphane Foucart, Stéphane Horel and Sylvain Laurel, the term serves as a "'Godwin point' of criticism of the intensive agriculture system" (Foucart et al. 2020, p. 27) and is part of a broad reactionary movement aimed at criticizing environmental movements as a whole.

Thus, we find within "agribashing" processes relating to a *formula*: the freezing and circulation in the public sphere of a single concept; the crystallization of social and political issues, here linked to the agricultural world, its models and its practices and, finally, a certain polemical character that precisely emphasizes "a conflictuality" (Simon 2015) specific to the ideological clashes which traverse the agricultural world.

1.3. Inclusion in the media agenda

The second part of our analysis focuses on the inclusion in the media agenda of the concept of "agribashing" . To do this, we created a corpus of articles from the national (NDP)[1] and regional (RDP)[2] daily press from October 2018, the date of its first appearance in the press, to October 2021, the closing date of our investigation. The corpus analysis provides us with interesting information about the peak of mentions of this term, its context and its spaces of media circulation.

At first glance, we see differences in treatment and media attention between the NDP and the RDP. While 1,356 articles from the RDP mention the term "agribashing", there are only 121 for the NDP over the same period. Analysis of the publication dates of the different articles mentioning the term "agribashing" allows us to observe the different periods during which the term was most common in the different media of our corpus (see Figure 1.1). In the regional press, the first article in our corpus to mention this term concerns an intrusion onto a farm which led the president of the Île-de-France Chamber of Agriculture to "[denounce] the daily 'agribashing' suffered by farmers" (Dutheil 2018), which corresponds well to a central claim of the agricultural movements at the time of the establishment of the Déméter task force. Conversely, in the national press, the first article from *Le Monde* mentioning the movement in December 2018 deals with fundamental opposition among farmers, and "agribashing" is associated with "measures taken by the government to try to support agriculture towards an ecological transition" (Prudhomme and Girard 2018), anchoring the ideological confrontation conveyed by this formula within the broader movement of transitions in the agricultural world. However, although the term appears as early as October 2018, the real "peak" of media attention only takes place between September 2019 and March 2020 when the majority of articles in our corpus were published for both the RDP (n = 893) and for the NDP (n = 78).

This peak of attention coincides with the victory of unions associated with the FNSEA (National Federation of Farmers' Unions) and JA (Young Farmers) in the professional elections of January 2019 and the launch of a joint campaign entitled "France, do you still want your farmers" (FNSEA 2019) as well as in the context of professional elections within Mutualité Sociale Agricole, the agricultural social insurance association.

1 The Europresse search engine includes some weekly news pages in the NDP category. We have chosen to retain them.

2 Europresse search, agribashing, NDP and RDP, from January 1, 2018 to October 31, 2021.

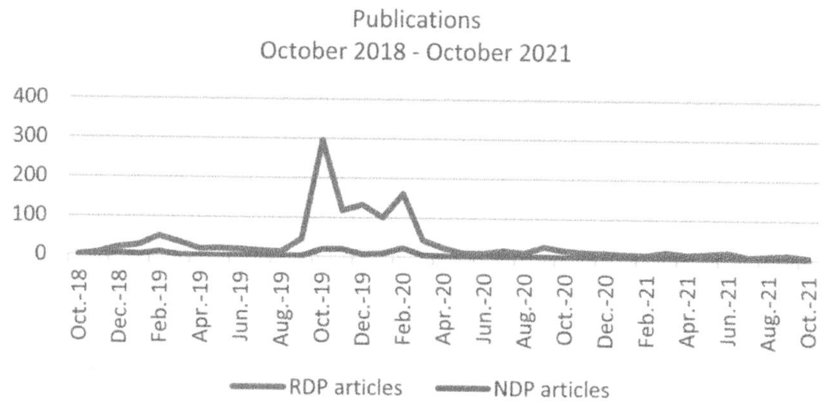

Figure 1.1. *Temporal distribution of articles. For a color version of this figure, see www.iste.co.uk/catellani/transitions.zip*

For comparison, we carried out a Google Trends analysis on the term "agribashing" which places the peak of interest and searches in November 2019, which might well correspond to interest linked to the appearance in the media of this term and to public curiosity about this new concept in circulation (see Figure 1.2).

In both cases, the number of publications mentioning the term "agribashing" decreases sharply after September 2020, never exceeding 20 mentions in the NDP. Thus, we can see that circulation of the term is directly linked to lobbying actions by the main French agricultural unions which, through mobilizing their electorates and their members, managed to put the term on the media agenda, even the public agenda, through various highly publicized and visible actions in the public space (demonstrations, blockades, etc.). For example, analysis of articles published in *Ouest-France* reveals more than 60 articles directly mentioning "agribashing" in their publication with very varied themes. If the question of anxieties in the agricultural world comes up very often – "FDSEA: concerns about "agribashing" or "Farmers worried about the rise of "agribashing" (Ouest-France 2019; Tymen 2019) – many publications discuss the measures put in place by the government to fight against it, or local initiatives which extend to education: "'Agribashing' explained to high school students" (Amiotte 2021). However, few articles directly refer to the oppositions that this term arouses. Ultimately, media attention seems to remain confined to local and regional issues, as the difference in the presence of the term between the major national dailies and the regional press is significant.

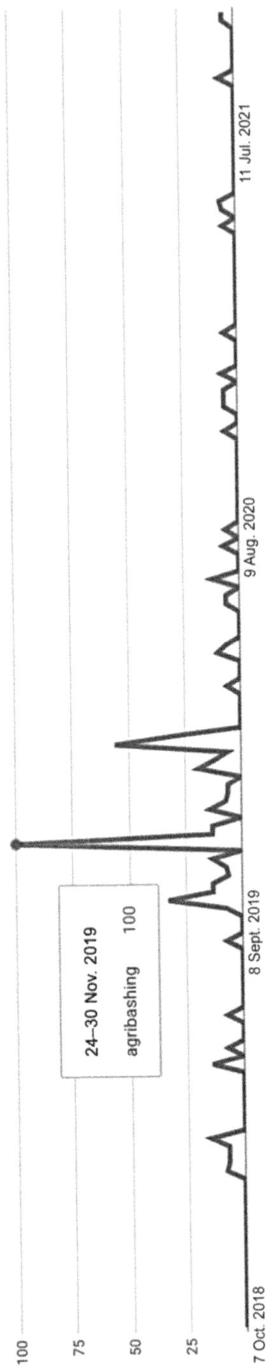

Figure 1.2. *Google trends analysis of agribashing*

Indeed, looking at the geographical distribution of the articles in our corpus, we see very strong disparities. Firstly, in terms of the national publications which are mainly interested in this question, *Le Monde* (n = 32), *Le Figaro* (n = 30) and *Libération* (n = 16) come first (see Figure 1.3). The other national titles in the corpus do not seem very interested in the question, which clearly reflects, over a period of two years, a lack of interest, or even the absence of an increase in the generality of the issues raised.

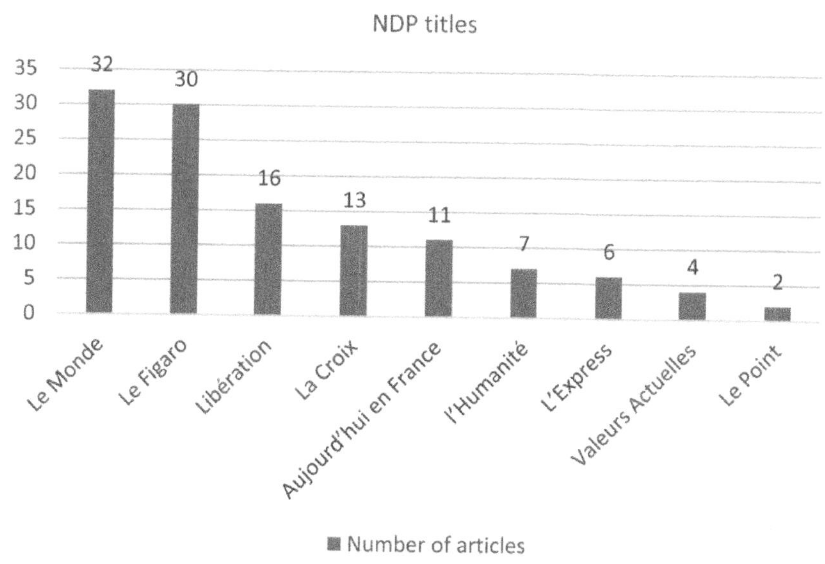

Figure 1.3. *Distribution of NDP publications*

In terms of the regional press (see Figure 1.4), we see a broad interest in publications from the West and South-West of France, with a majority of articles devoted to the issue by *Ouest-France* (n = 326) followed by *Sud-Ouest*, print and Web (n = 227), *Le Télégramme* (n = 107) and *La Nouvelle République du Centre-Ouest* (n = 104). In the East and South-East, it is essentially *L'Est Républicain* (n = 105) and *Le Progrès de Lyon* (n = 91) that deal with this theme, far ahead of other regional titles.

These different spatial distributions also seem to reflect differences of interest in the subject according to agricultural specializations. Indeed, the Western and Central-West regions, which are mainly oriented towards livestock farming (Agreste 2020), are those where the subject is most covered in the regional press. This interest is consistent with the agricultural world's concerns around the issue of "intrusions" into farms and anti-speciesist activism.

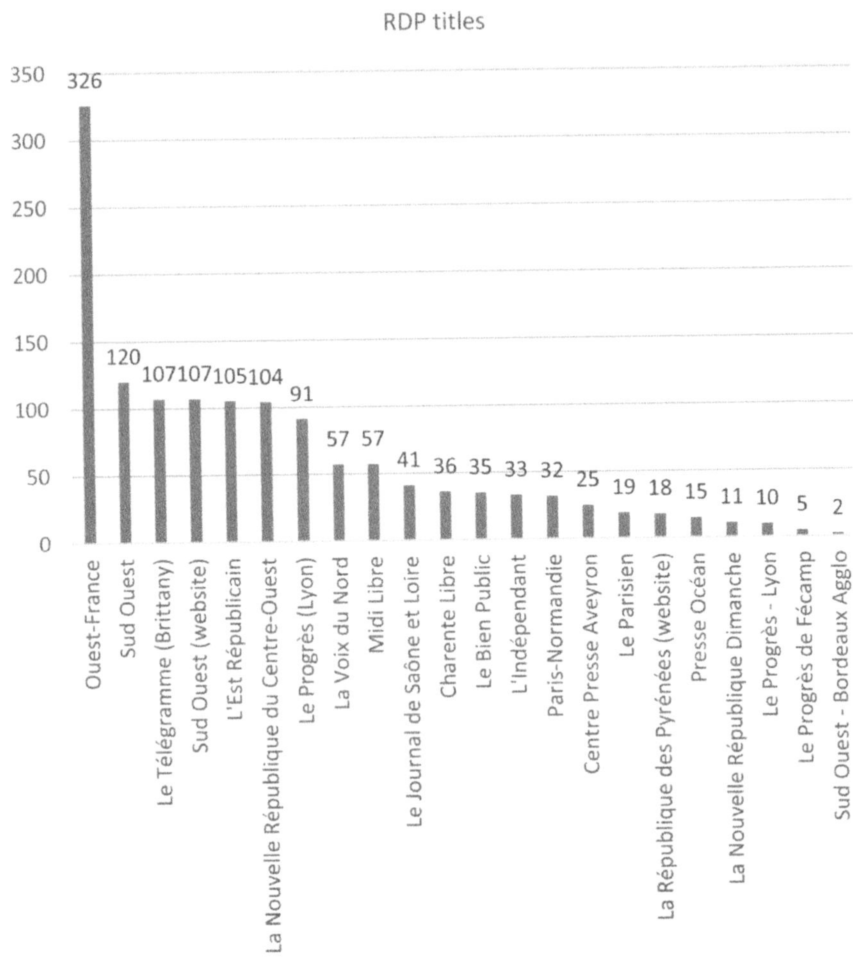

Figure 1.4. *Distribution of RDP publications*

1.4. Disqualification of dissenting voices

Analysis of the spatio-temporal circulation of the term "agribashing", as well as analysis of the context of this circulation, allowed us to understand the strong links between the inclusion of this term on the media agenda and the political interests of certain actors in the agricultural world. A thematic and discursive analysis of the articles, supplemented by the reading of documents linked to the Déméter task force, also allows us to better understand the ideological issues that make this concept a *formula* in its own right.

We performed an analysis of the main terms present in the two corpora of NDP and RDP. Using the AntConc software (Anthony 2005), we observed the frequency of terms and their co-occurrences, and we produced "word clouds" to visualize the terms most frequently associated with "agribashing" in both corpora.

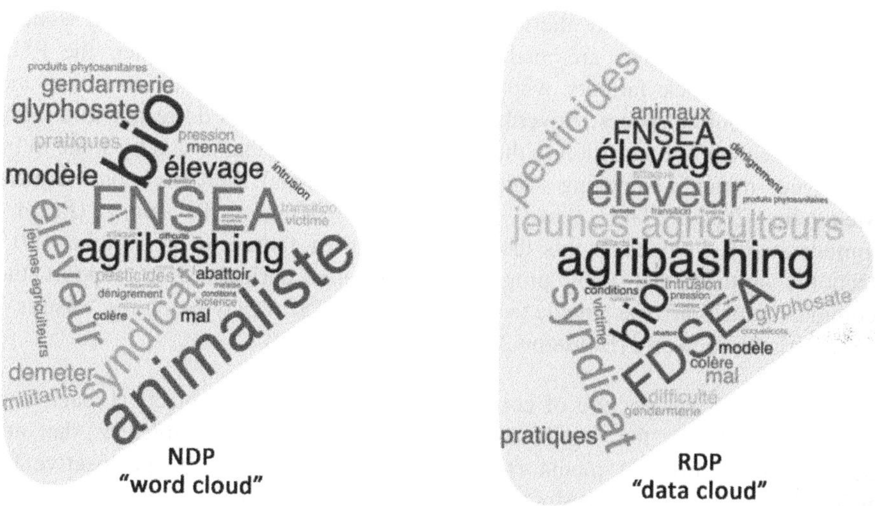

Figure 1.5. *"Word cloud" for NDP and RDP in French. For a color version of this figure, see www.iste.co.uk/catellani/transitions.zip*

As can be seen in the two "clouds", the presence of agricultural unions (FNSEA, FDSEA, Young Farmers) strongly dominates the corpus, with, however, a more "local" anchoring in the RDP corpus where the different local sections of the FNSEA, the FDSEAs, are the most commonly mentioned. Although the question of livestock farming is present in both cases with its lexical field, we notice an important focus on the figure of the "livestock farmer" who is foregrounded. Another point of convergence between the two corpora is the question of pesticides, which often arises as a stumbling block specific to "agribashing". We may nevertheless note that although the generic term "pesticides" is used in both corpora, the RDP titles use the positive name "phytosanitary products" much more clearly than the national titles. In this respect, we note the interesting presence of "glyphosate", which often attracts public attention and is a marker of current media polemics around agricultural models in France (Allard-Huver 2021, 2022).

Finally, within the articles, the question of different agricultural practices is at the heart of the debate with the opposition between the so-called "conventional", "intensive" or "agro-industrial" models, on the one hand, and the "organic" model, on the other. However, the question of ideological confrontation around the figure of animal rights movements is much more present in the NDP corpus, with a debate around the very nature and mission of the Déméter task force (Gendarmerie nationale 2019). The media framing is then constructed around a different perception: on the one hand, the NDP presents those who fight for animal welfare and for whom intrusions are justified, the "animalists"; on the other, the RDP focuses on livestock farmers who are "victims" of attacks and intrusions and denigrated in public spaces. Nevertheless, some of the articles do testify to a more global movement of opposition to the ecological transition that is said by some to be a "forced march" imposing "unacceptable additional burdens" on farmers (Menvielle 2018). The actors of the Confédération Paysanne (Peasant Confederation) even see in this the reasons for their failure in the October 2019 elections, arguing that their program "to develop agriculture to support the ecological transition" was unable to cope with the FNSEA playing on the "distress" of the agricultural world (Boussion 2019).

Thus, from a perspective of communication on sensitive subjects (Libaert and Allard-Huver 2014), the "agribashing" formula serves several interests that are found in these different media framings Firstly, it reverses the perspective by creating a story where livestock farmers become the victims of a systematic denigration of their practices, which makes it possible to avoid the axiological debate concerning the condition of animals. Furthermore, it operates a form of syllogism where any criticism of the intensive agricultural model and its chemical avatars (pesticides, glyphosate, non-treatment zones, etc.) is considered as a form of "stigmatization of agricultural activities" (Seine-Maritime Department 2020). Finally, in terms of controlling the framing of the public sphere, this formula seems to contribute to muzzling, with the help of public authorities, any form of investigation or communicational action by civil society actors engaged in the agroecological transition (Astier and Rimbert 2020; Chapelle 2020; Pollinis 2021).

1.5. Conclusion

From a semiotic and discursive point of view, the term "agribashing" is a "formula" (Krieg-Planque 2009; Simon 2015) at the heart of a broader debate, a global controversy around the transition of the agricultural model that crystallizes the oppositions between the actors (Pascual Espuny 2014, 2017). Though the *formula* appearing in 2018 then circulating in the media until 2020 is, at first glance, the expression of a certain unease in the agricultural world, it is just as much the politicization of an ideological argument disqualifying any dissenting voices.

Thus, this work is to be seen in the context of proposed reflections on the production, circulation and acceptance of discourses around the agricultural transition and the controversies they may arouse (Aulagnier and Goulet 2017). Although we note the slow disappearance of the term in the media space, we nevertheless observe its return in the specific space of social networks where it is still widely used by some radical defenders of the productivist model. This polytopia of the concept (Jeanneret 1998) also confirms that beyond its deeply polemical nature, the "half-life" of this concept can only be seen as consequent to a lobbying campaign by certain actors in the agricultural world. Furthermore, this campaign allowed the creation of the "Déméter" task force which in turn sparked significant controversies. The creation of this tool is part of a complex context in which certain actors in civil society criticize the appropriateness of the State using the police force to decide, ideologically, which discourses around the ecological transition have the right to be heard, or not, just as it reflects the changes in operation within the agricultural world around a deeper debate on the models of the agroecological transition to come.

1.6. References

AFP and Le Monde (2021). Suicide chez les agriculteurs : le gouvernement lance une "mobilisation collective". *Le Monde.*

Agreste (2020). Carte interactive – spécialisation territoriale de la production agricole en 2020. Agreste, la statistique agricole [Online]. Available at: https://www.agreste.agriculture.gouv.fr/agreste-web/disaron/Carte-RA2020-otex/detail/ [Accessed 7 June 2022].

Agriculture et Environnement (2016). Comment sortir de l'agriculture bashing ? [Online]. Available at: https://www.agriculture-environnement.fr/2016/03/08/comment-sortir-de-lagriculture-bashing.

Allard-Huver, F. (2021). Agnotologie, transparence et fake news : stratégies de guerre informationnelle autour du glyphosate. *Revue internationale d'intelligence économique*, 13, 53–75.

Allard-Huver, F. (2022). Glyphosate, la "guerre des urines" a bien eu lieu. *Les Enjeux de l'information et de la communication*, 22, 11–24.

Amiotte, S. (2021). L'agribashing expliqué aux lycéens. *Ouest-France Chateaubriant*, 11.

ANIA and TNS-Sofres (2013). Résultats du baromètre ania-TNS Sofres. *Les Français et l'alimentation.*

Anthony, L. (2005). AntConc: A learner and classroom friendly, multi-platform corpus analysis toolkit. In *Proceedings of IWLeL 2004: An Interactive Workshop on Language e-Learning*. Waseda University, Tokyo, 7–13.

Astier, M. and Rimbert, I. (2020). Déméter, la cellule de la gendarmerie qui surveille les opposants à l'agriculture productiviste. *Reporterre, le quotidien de l'écologie*. [Online]. Available at: https://reporterre.net/Demeter-la-cellule-de-la-gendarmerie-qui-surveille-les-opposants-a-l-agriculture [Accessed 8 May 2021].

Aulagnier, A. and Goulet, F. (2017). Des technologies controversées et de leurs alternatives. Le cas des pesticides agricoles en France. *Sociologie du travail*, 59. doi: 10.4000/sdt.840.

Bernard, F. (2018). Les SIC et l'"anthropocène" : une rencontre épistémique contre nature ? *Les Cahiers du numérique. Pluralisme épistémologique et conceptuel en information-communication*, 14, 31–66.

Boussion, X. (2019). La FNSEA reprend la chambre d'agriculture. *Presse Océan PO Saint-Nazaire Presqu'île*, 5.

BVA and Crédit Agricole (2021). Baromètre BVA-Crédit Agricole – #Agridemain – sur l'image de l'agriculture auprès des Français [Online]. Available at: https://agridemain.fr/wp-content/uploads/2021/05/BVA-Sondage-Image-de-lagriculture-aupres-des-Francais-en-2021-12-Mai-2021.pdf.

Catellani, A., Pascual Espuny, C., Malibabo, L.P., Vigouroux, B.J. (2018). Post-normal science and environmental communication: Talking about risk, uncertainty and controversies. *7th European Communication Conference – Centers and Peripheries: Communication, Research, Translation – ECREA Congress*, Lugano.

Chapelle, S. (2020). Cellule de renseignement Demeter : quand la gendarmerie se met au service de l'agro-industrie. *Basta!* [Online]. Available at: https://www.bastamag.net/cellule-de-renseignement-Demeter-lutte-agribashing-quand-la-gendarmerie-se-met-au-service-de-la-FNSEA [Accessed 8 May 2021].

Chardon, O., Jauneau, Y., Vidalenc, J. (2020). Les agriculteurs : de moins en moins nombreux et de plus en plus d'hommes. *INSEE – Focus*, 212.

Delame, N. (2021). Revenus agricoles et non agricoles des agriculteurs de 2003 à 2016. *Économie rurale*, 378, 77–95. doi: 10.4000/economierurale.9422.

Département Seine-Maritime (2020). Observatoire départemental de l'agribashing [Online]. Available at: https://www.seine-maritime.gouv.fr/Actualites/Observatoire-departemental-de-l-agribashing [Accessed 8 May 2021].

Dutheil, M.-C. (2018). Des militants volent dans les plumes d'un élevage de poules. *Le Parisien*.

FNSEA (2019). France, veux-tu encore de tes paysans ? [Online]. Available at: https://www.fnsea.fr/communiques-de-presse/france-veux-tu-encore-de-tes-paysans/ [Accessed 7 June 2022].

Foucart, S., Horel, S., Laurens, S. (2020). *Les gardiens de la raison : enquête sur la désinformation scientifique*. La Découverte, Paris.

Fougier, E. (2018). Le monde agricole face au défi de l'agribashing. Report, FNSEA, Paris [Online]. Available at: https://www.fnsea-bassin-parisien.fr/wp-content/uploads/2018/09/Synth%C3%A8se_Rapport-agribashing_E.-Fougier.pdf [Accessed 7 June 2022].

Fougier, E. (2020). *Malaise à la ferme : enquête sur l'agribashing*. Éditions Marie B, Clichy.

FrenchFood Capital (2018). Les Français et l'alimentation : exigence et vigilance sur la composition et la qualité des produits [Online]. Available at: https://frenchfoodcapital.com/les-francais-et-lalimentation-exigence-et-vigilance-sur-la-composition-et-la-qualite-des-produits/ [Accessed 8 May 2021].

Gendarmerie nationale (2019). "Déméter", une cellule nationale pour lutter contre l'agribashing et les intrusions dans les exploitations agricoles [Online]. Available at: https://agriculture.gouv.fr/demeter-une-cellule-nationale-pour-lutter-contre-lagribashing-et-les-intrusions-dans-les [Accessed 8 May 2021].

IFOP and Ouest-France (2022). Baromètre d'image des agriculteurs. *Ifop pour Dimanche Ouest France.*

Jeanneret, Y. (1998). L'affaire Sokal : comprendre la trivialité. *Communication & Langages*, 118, 13–26. doi: 10.3406/colan.1998.2879.

Krieg-Planque, A. (2009). *La notion de formule en analyse du discours : cadre théorique et méthodologique*. Presses Universitaires de Franche-Comté, Besançon.

Le Hesran, C. (2020). "France, veux-tu encore de tes paysans ?" : les agriculteurs se mobilisent mardi 8 octobre sur les routes. *France 3 Nouvelle-Aquitaine.*

Libaert, T. and Allard-Huver, F. (2014). La communication sur les sujets sensibles au prisme des sciences de l'information et de la communication. *Communiquer. Revue de communication sociale et publique*, 11, 81–100. doi: 10.4000/communiquer.574.

Menvielle, D. (2018). Les agriculteurs lancent un ultimatum au gouvernement. *Le Progrès.*

Ministère de l'Intérieur (2019). Présentation de "DEMETER", la cellule nationale de suivi des atteintes au monde agricole [Online]. Available at: http://www.interieur.gouv.fr/Archives/Archives-ministres-de-l-Interieur/Archives-Christophe-Castaner/Dossiers-de-presse/Presentation-de-DEMETER-la-cellule-nationale-de-suivi-des-atteintes-au-monde-agricole [Accessed 8 May 2021].

MSA (2020). La MSA agit pour prévenir le suicide en agriculture. Press release, Bobigny.

Ouest-France (2019). FDSEA : des inquiétudes face à l'agribashing. *Ouest-France Centre-Finistère*, 11.

Pascual Espuny, C. (2014). La société civile, de l'alerte à la controverse médiatisée. *Communication et organisation*, 115–126. doi: 10.4000/communicationorganisation.4531.

Pascual Espuny, C. (2017). Communication environnementale et communication des organisations. Logiques de publicisation, de circulation et de cristallisation. HDR Thesis, Aix-Marseille Université.

Pollinis (2021). Cellule Déméter, le bras armé de la FNSEA qui empêche de repenser l'agriculture [Online]. Available at: https://www.pollinis.org/publications/cellule-demeter-linquietant-bras-arme-au-service-de-lagriculture-conventionnelle/ [Accessed 8 May 2021].

Prudhomme, C. and Girard, L. (2018). La pression monte chez les agriculteurs. *Le Monde*.

Rass, G. (2020). Les agriculteurs sont dépossédés par les associations. *Paysans & société*, 379, 9–19. doi: 10.3917/pes.379.0009.

Simon, J. (2015). Formule. *Publictionnaire. Dictionnaire encyclopédique et critique des publics* [Online]. Available at: http://publictionnaire.humanum.fr/notice/formule/.

Tymen, C. (2019). Les agriculteurs inquiets de la montée de l'agribashing. *Ouest-France*.

2

A "Showcase Agriculture" Around Cities? Urban Food Policies, a Communication Tool for Farmers' Organizations

The Milan Urban Food Policy Pact is a network of cities committed to improving the sustainability of their food systems. This network, created in 2015, brought together more than 270 signatories in 2023, witnesses to the dynamism of cities regarding urban food planning (Moragues-Faus and Morgan 2015). This has been the case particularly in France since the 2000s (Brand 2017). The French urban food policies have the feature of being strongly connected to agricultural issues (Perrin and Soulard 2017). They promote alternative agricultural development by encouraging organic farming, short food supply chains or crop diversification (Pahun 2020).

These farming practices represent an alternative to the productivist, chemical and export-oriented model of agriculture, a model still widely defended by the majority representatives of the sector in France, namely the FNSEA[1], its local sections and the Chambers of Agriculture, of which they direct over 95%. These last two actors, whom we call here "local agricultural elites" (Sarrazin 2014), are traditionally disinclined to consider the multidimensional challenges of agriculture, whether pertaining to the environment (Szarka 2000) or emerging food-related issues (Hobeika 2020; Mayance 2020). However, it is noteworthy that they emerge as key partners in urban food policies in France (Pahun 2020). We might therefore wonder why these local agricultural elites collaborate in urban food policies, when their

Chapter written by Jeanne PAHUN.

1 *Fédération nationale des syndicats d'exploitants agricoles*, National Federation of Farmers' Unions.

political vision of agricultural development seems to be at first sight to be highly divergent.

To explain this contradiction, we draw upon three case studies concerning the construction and implementation of food policies in the cities and metropolises of Rennes, Strasbourg and Montpellier. Between 2018 and 2019, we conducted a series of interviews with 26 key players in the agricultural aspects of these food policies, including elected officials and agents of the relevant urban centers (14), elected officials and agents of the partnering Chambers of Agriculture (7) and elected officials and agents of alternative agricultural advisory structures associated with these food policies (5). This interview-based investigative effort was complemented by a study of the available gray literature on the policies studied: action plans, budgets, institutional magazines and communication brochures. This investigation was conducted as part of a doctoral thesis on territorial food policies in France (Pahun 2020).

In this chapter, we first describe the actions carried out in the agricultural sector within the framework of the three urban food policies studied and the relationships maintained by municipal or metropolitan teams with local agricultural elites around this topic. We then analyze the communication issues that arise in these collaborations. Finally, we conclude on the symbolic and strategic dimension that these partnerships with cities and their food policies represent for local agricultural elites and majority agricultural representatives in general.

2.1. The agricultural aspects of urban food policies

The three urban food policies studied are recent policies, initiated between 2008 and 2017. These policies encompass a range of diverse actions such as the renovation of farmers' shops in city centers, the organization of cultural events around food, the reduction of waste in school canteens or the cataloging of local producers engaged in short food supply chains. Among the different areas of intervention, the agricultural sector holds a prominent position within these food policies.

The sustainable food plan for the city of Rennes, initiated in 2017, initially focused on improving collective catering in the city's primary schools. The procurement of school canteens was then seen as a lever for action in the agricultural sector. Because, as recalled by the elected official leading the city's food policy:

> Unlike the metropolis which has real agricultural competence, we [the city of Rennes] do not have an agricultural department, clearly not. But for us, our lever for action is the act of purchasing food for school

> catering. […] I want the city's purchasing policy to be a key lever for transforming agriculture towards more sustainable production methods[2].

Thus, to further control the supply of its primary school canteens (for which it has legal responsibility) and to increase its share of products from organic farming, the municipality of Rennes has developed a new public market. This market is specifically intended for farmers located in the drinking water catchments supplying the city and engaged in an IDEA approach (*indicateurs de durabilité des exploitations agricoles*, sustainability indicators for agricultural operations) to improve their agricultural practices (territorial anchoring, autonomy, crop diversification, animal welfare, etc.). This market, orchestrated in collaboration with the Rennes Bassin water community, brought together 20 producers in the Rennes region in 2019.

In Strasbourg, the "Strategy for the development of a sustainable and innovative local agriculture responding to the food needs of the population" was initiated in 2008. Its primary objective was to develop short food supply chains within the city and metropolis of Strasbourg. The first action implemented within this policy was the assessment of agricultural operations present in the territory. Faced with the observation that local agriculture was more than 70% oriented towards field crops and export, an action plan for the transformation of peri-urban agriculture was enacted. This aimed, on the one hand, to preserve agricultural land in the Strasbourg metropolitan area and, on the other hand, to steer farms towards more sustainable agriculture, encompassing greater crop diversification, organic farming and the development of short food supply chains.

The "Montpellier metropolitan agroecological and food policy" was voted on for the first time in 2015. Its agricultural component consisted of several flagship actions, including the installation of farmers on metropolis-owned land. These spaces, called "agriparks", were intended to reconcile urban functions (educational, recreational, food-related, environmental) with agricultural functions (economic and productive). One of the metropolis's emblematic agripark projects was the installation of a collective of five market gardeners working outside a farming family structure and "mobilized to promote a nourishing agroecology"[3]. The farmers were selected through a call for projects organized by the metropolitan food policy team. The specifications of this call for projects were precise: the productions had to be labeled as organic farming, half of the commercial outlets had to be in short food

2 Interview with the deputy mayor of the City of Rennes in charge of the public procurement delegation and initiator of the city's sustainable food plan, Rennes, March 16, 2018.

3 Call for projects: "Consolidation of an agroecological food-producing farms network. Domaine de la Condamine's allocation", Montpellier Méditerranée Métropole, 2015.

supply chains, and farm visits for schools, leisure centers or individuals had to be integrated into the operation.

The problematization of food issues enacted by the three municipal teams studied therefore widely integrates agricultural issues. Another common point of this problematization is the desire of urban teams to address themselves to the professional agriculture sector. Indeed, forms of intra-urban agriculture (shared gardens, vertical farms or container farming) were never mentioned during our interviews with food policy representatives, nor do they appear in the agricultural aspects of food policy. In section 2.2, we will study the partnerships with actors in the agricultural world set up by urban teams to support this political desire to engage with the professional agricultural sector from which urban governments had become increasingly estranged (Daviron et al. 2017).

2.2. Collaborations between local agricultural elites and municipal or metropolitan teams

The relationships established by urban food policy teams with actors in the agricultural sector vary across the policies studied. In Rennes, the city did not previously have a charter linking it to agricultural partners. When developing its sustainable food plan, it involved the Chamber of Agriculture. But the day before submitting an application to a national call for projects aimed at partly financing the policy, the Chamber of Agriculture withdrew from the partnership, following a disagreement on the notion of sustainable agriculture, before finally returning a few months later. The city teams therefore do not have a stable political partnership with the Chamber of Agriculture, and we see several back-and-forths between these two entities, the local Chamber of Agriculture navigating between active participation, negotiation and partnership breakdown.

In Strasbourg, the food strategy is an opportunity to establish an initial partnership between the municipal team associated with metropolitan agents and the local Chamber of Agriculture. The Chamber of Agriculture is even the first partner to whom the elected representative in charge of the food policy turned. A charter internally entitled "Evolution of territories towards sustainable agriculture" was signed between these two entities at the launch of the food strategy. This charter was the site of negotiations and compromises between the two institutions (on the issue of land use, the environment or organic farming, for example). The Chamber of Agriculture is therefore the main partner of Strasbourg's food policy, and relations between the two entities are robust.

When the elected representative in charge of Montpellier's agroecological and food policy came to power, she decided to suspend the convention that bound the metropolis to the local Chamber of Agriculture and to rewrite it entirely. She explains:

> When I arrived, this convention was almost entirely written by the Chamber of Agriculture and the Chamber was at the same time the principal, the designer, and the service provider. The agroecological and food policy had to be integrated into the chamber's policy. [...] We didn't want it to be an offshoot of their policy[4].

Despite the rewriting of this convention, which might have led to tensions between the two institutions, the resulting collaborations between agents of the chamber and agents of the metropolis were dense and relatively smooth during the initial implementation of the policy. Furthermore, it should be noted that, in the cases of Montpellier and Strasbourg, these partnership agreements with the local Chambers of Agriculture were backed up by partnership agreements with alternative agricultural development structures, the Inpact[5] network in Montpellier and Opaba[6] in Strasbourg.

While relations between local agricultural elites and municipal teams appear more conflictual in the case of Rennes' food policy, the question is why, in the case of Montpellier and Strasbourg, the Chambers of Agriculture and their union leaders have volunteered to support the food policies of these two metropolises which notably aim to develop an agricultural alternative to the model predominantly supported by this very network of Chambers of Agriculture.

2.3. Communication issues at the heart of collaborations between local agricultural elites and urban centers

Several elements support the thesis that the communication issues represented by the deployment of urban food policies for local agricultural elites have been one of the driving forces behind their collaboration in these policies.

The first element supporting this thesis is the figure of "ambassador farmers" mentioned by several elected representatives of Chambers of Agriculture who were surveyed. For example, in the case of Strasbourg, the alliance between the Chamber and the metropolitan team is partly due to the appointment of a new president of the

---

4 Interview with the vice-president of agroecology and food for the Montpellier Metropolitan Area, Montpellier, October 26, 2018.

5 Inpact, *Initiatives pour une agriculture citoyenne et territoriale* (Initiatives for Citizen-based and Territorial Agriculture).

6 Opaba, *Organisation professionnelle de l'agriculture biologique en Alsace* (Professional Organization for Organic Agriculture in Alsace).

Chamber, whose mandate project included strengthening partnerships with local territorial authorities and supporting short food supply chains. This project was partly justified by the desire to deploy "ambassador farmers" around cities, as explained in this interview excerpt:

> [The president] has a very sector-oriented approach, he comes from that culture, on the other hand when he arrived [at the head of the Chamber] he said: "Today's consumer expectations are based on short food supply chains and we are in a department where there are 400 inhabitants/km^2, so we cannot say we don't care about short food supply chains". He immediately saw the role of these short food supply chains, because a farmer who sells in short food supply chains is an ambassador for the agricultural world[7].

Thus, to encourage short food supply chains is to encourage the installation of farmers who will be in direct contact with the population. And in this case, the population of urban centers, places of concentration of the population, on the one hand, and formation of opinion, on the other. This idea is expressed in a second excerpt from an interview with this head of services at the Alsace Chamber of Agriculture:

> Even if there are only 10,000 hectares in the Eurometropolis, there are also a lot of people who talk, who think. Ideas that circulate across the entire Alsace region start from things that are observed here, around the cities. So, I fully agree that we should work together [with the city and the metropolis of Strasbourg][8].

The figure of "ambassador farmers" was therefore one of the driving forces behind the partnership established between the Alsace Chamber of Agriculture and the urban community of Strasbourg. In the case of Montpellier, the president of the Occitanie Regional Chamber of Agriculture expressed the same idea, that of building "positive communication" about the agricultural sector through the development of peri-urban agriculture oriented towards the new expectations of urban dwellers concerning their food and agriculture. He stated thus during an interview: "The goal is to have farmers around the cities who can welcome people, we want to adopt positive communication"[9].

7 Interview with a head of services of the Alsace Departmental Chamber of Agriculture, Schiltigheim, May 24, 2018.

8 Ibid.

9 Interview with the president of the Occitanie Regional Chamber of Agriculture, Lattes, December 20, 2018.

Urban food policies are thus an opportunity to deploy "showcase agriculture" around cities, a term we use to describe these farms serving as "positive communication" for representatives of the agricultural world. The type of rather alternative farms that may be part of this peri-urban agriculture (organic, diversified, run by neo-rurals, open to the public, etc.) is here endorsed by the majority representatives of the agricultural world, as it improves the image of the sector among urban dwellers.

It should be noted that this type of agriculture is more often referred to as niche agriculture by the majority representatives of the profession than as a model for the future development of agriculture. Thus, beyond agreement in principle on the deployment of this alternative agriculture around cities, certain agricultural leaders interviewed continued to give it no credit, as evidenced by the following interview excerpt:

> Agripark X of the Montpellier metropolis, I am not sure that it's a success. […] I am also the president of the Safer[10] of Hérault county, we handed over land to them for these things, but the guys, they don't stay there.
>
> Interviewer: Yes, but it's the principle of a farm incubator.
>
> Yes, but then they don't settle. I went there for the inauguration with three boxes of twisted carrots you can't… well, there are some customers who go there. But later we see all these farmers in "agridiff" [aid program for farmers in difficulty]. Initially it works a little because they are employees, so they receive social assistance[11].

This contempt displayed towards the farmers settled around Montpellier under the aegis of the metropolitan food policy contrasts with the *laissez-faire* and collaboration of the local agricultural elites in this urban food policy.

2.4. Conclusion

Even if the type of agriculture supported within the food policies of the Strasbourg and Montpellier metropolitan areas is far removed from the agricultural development mainly promoted by the Chambers of Agriculture, the latter have a strategic interest in taking part and collaborating with it. Apart from diversifying and

10 Safer, *Société d'aménagement foncier et d'établissement rural* (Land Management and Rural Establishment Agencies).

11 Interview with the president of the Occitanie Regional Chamber of Agriculture, Lattes, December 20, 2018.

stabilizing their sources of funding, the partnerships established with urban administrations are part of a strategy to improve the symbolic capital of the agricultural world, deploying "ambassador farmers" around places where public opinion is shaped, conveying a positive image of the sector to urban residents.

Urban food policies thus allow local agricultural elites to contribute to the construction of a new image of the agricultural world, closer to new citizen expectations, through the establishment of "showcase agriculture" around cities. In this sense, urban food policies are also "image policies" (Le Bart 1999) as they convey positive mental representations of the agricultural sector, such as these peri-urban agroecological market gardening farms, featuring organic farming and short food supply chains.

The establishment of "showcase agriculture" around cities is also part of a logic of showing a change in the agricultural sector which is not occurring elsewhere. There is then a risk of focusing, as a researcher or as a citizen, on these urban actions promoting sustainable agriculture, which are sometimes presented as the future of the agricultural transition when they are only a showcase for it.

2.5. References

Brand, C. (2017). The French urban food issue emergence. *Bollettino della Società Geografica Italiana*, 10(13), 65–76.

Daviron, B., Perrin, C., Soulard, C.T. (2017). Histoire des politiques alimentaires urbaines en Europe, de la ville antique à la ville industrielle. In *Construire des politiques alimentaires urbaines*, Brand, C., Bricas, N., Conaré, D., Daviron, B., Debru, J., Michel, L., Soulard, C.T. (eds). Quae, Versailles, 43–65.

Hobeika, A. (2020). Résistances de la FNSEA aux problématisations environnementales et alimentaires de l'agriculture. In *Quand l'alimentation se fait politique(s)*, Fouilleux, E. and Michel, L. (eds). Presses Universitaires de Rennes, Rennes, 125–142.

Le Bart, C. (1999). Les politiques d'image. Entre marketing territorial et identité locale. In *Les nouvelles politiques locales. Dynamiques de l'action publique*, Faure, A. and Balme, R. (eds). Presses de la Fondation Nationale des Sciences Politiques, Paris, 415–427.

Mayance, P. (2020). Le cas de l'agriculture raisonnée en France. In *Quand l'alimentation se fait politique(s)*, Fouilleux, E. and Michel, L. (eds). Presses Universitaires de Rennes, Rennes, 143–160.

Moragues-Faus, A. and Morgan, K. (2015). Reframing the foodscape: The emergent world of urban food policy. *Environment and Planning A: Economy and Space*, 47(7), 1558–1573.

Pahun, J. (2020). L'agriculture face aux politiques alimentaires : une analyse comparée dans trois régions françaises. PhD Thesis, Université Paris-Est, Paris.

Perrin, C. and Soulard, C.T. (2017). Introduction. L'agriculture dans le système alimentaire urbain : continuités et innovations. *Natures sciences sociétés*, 25(1), 3–6. doi: 10.1051/nss/2017012.

Sarrazin, F. (2014). *Les élites agricoles et rurales. Concurrences et complémentarités des projets*. Presses Universitaires de Rennes, Rennes.

Szarka, J. (2000). Environmental policy and neo-corporatism in France. *Environmental Politics*, 9(3), 89–108.

3

Local Conflicts Around Creative Development Projects

Since the 2000s, local territories have proved to be spaces of tension between two paradigms of culture and creation. On the one hand, particularly in most of the metropolises of Western Europe including France, phenomena of territorialization of actors in the cultural, creative and digital industries have increased, jointly supported by public institutions (local authorities, economic chambers, professional federations) and these industrial actors, sometimes with the legitimizing reinforcement of recognized experts and consultants. Whether they produce cultural goods or integrate dimensions and practices relating to the creative economy into their process, these industries have the common factor that they consider "creativity" and esthetic attributes to be the determining factors in differentiating their products and services and their ability to create value in markets (Vincent and Wunderle 2012). This phenomenon is seen in most industrial sectors, particularly in the form of proposals for consumer "experiences". Activities perceived as innovative are established in the territories, whether by the relative originality of their offer or by mobilization of new technologies, in particular digital technologies, which, in addition to their effective contribution to economic projects, confer upon them the symbolic capital of modernity and inclusion in global economic dynamics.

The leisure, entertainment and tourism economies are of course involved, but this phenomenon also affects economies not directly related to creative activities, such as e-commerce, logistics and mobility. The ideological dimension of this phenomenon is essential: the belief in the capacity of economies perceived as "creative" to boost territorial socio-economic development (Landry 2000; Higgs et al. 2008) constitutes a dominant frame of thought for the construction of public decision-making. On the other hand, and significantly since 2018, a growing number

---

Chapter written by Bruno LEFÈVRE.

of actors adept in the discourse of representatives of "civil society" and "citizen collectives", associated with activist associative networks, oppose these projects and defend a paradigm of development focused on the search for new forms of organization and production of "public goods", even "commons". The logic of distinction on the markets is confronted here by a holistic framework of development thinking, integrating a broad spectrum of local and international issues. One of the objectives of these adversarial dynamics lies in the establishment of forms of governance that would allow, at different scales, the direction of territorial development by integrating the cultural, societal, environmental and socio-economic impacts of decisions.

These tensions, observable in France in relation to several hundred local projects[1], challenge not only models of territorial development and the role of the cultural, creative and digital industries in these strategies (Bouquillion 2012), but also the modalities of management of ideological conflicts in decision-making processes by local public decision-makers (Harvey 2011).

3.1. Local public spaces that "repoliticize" public decision-making

Here, we hypothesize a contemporary phenomenon of "repoliticization" of local actors because of such conflicts related to territorial development and planning and which challenge institutional narratives that consider culture and creativity as resources for distinction and attractiveness. More specifically, we seek to characterize the processes through which reactions, partly sensitive and intimate, to local projects evolve into individual and collective thought and action, beyond the local level, to federate in national and international networks and dynamics, advocating for integration of social and environmental issues into the frameworks of institutional decision-making (Cefaï 2007). This process seems to also find its justification in a feeling, widely shared among citizens, of a paradoxical form of depoliticization of the sphere of public decision-making, and therefore of the political sphere; faced with forms of rationality and procedures related to decision-making, forms of political thought and conflict management are put forward which are claimed to be absent from institutional practices. Our observations and analyses thus focus on the potential for the emergence of new forms of public spaces contributing, through power relations and within a hegemonic capitalist economy, to offering frames of thought for local action integrating major international issues that have hitherto been little present.

---

1 See the survey conducted by Reporterre: https://lutteslocales.gogocarto.fr/annuaire#/carte/@46.33,2.50,6z?cat=all.

Project	Industrial sector	Territory concerned	Announcement date	Land acquired
Alibaba	E-commerce	Liège (Wallonia – Belgium)	2018	22 ha
	E-commerce	Belin-Béliet (Gironde)	2018	19 ha
	E-commerce	Illies (Nord)	2018	32 ha
Amazon	E-commerce	Dambach-la-Ville (Alsace)	2019	17 ha
	E-commerce	Einsisheim (Alsace)	2019	19 ha
	E-commerce	Briec (Loire-Atlantique)	2019	6 ha
	E-commerce	Augny (Lorraine)	2019	19 ha
Europacity	Leisure business park	Gonesse (Val d'Oise)	Abandoned 2019	80 ha
Imagiland	Amusement park	Angoulême	2019	12 ha
Melofolia	Amusement park	Coussac-Bonneval (Haute-Vienne)	2019	26/141 ha
Saint-Sauveur	Creative district	Lille (Nord)	2013	23 ha
Tropicalia	Amusement park	Berck-sur-Mer (Pas-de-Calais)	2018	10 ha
Studios Occitanie Méditerranée	Media and cultural industries complex	Béziers (Hérault)	2019	78 ha
D-Day Land	Amusement park	Manche	2020	35 ha

Table 3.1. *Main characteristics of the development projects observed*

To do this, in 2020 and 2021, we contrastively observed approximately 15 French and Belgian local economic development projects that were subject to contestation by local civil society and activist networks (see Table 3.1). These projects involve, on the one hand, infrastructures linked to e-commerce and digital industry (including Amazon and Alibaba logistics centers) and, on the other hand, cultural facilities or amusement parks with cultural or creative themes (including

Imagiland, Europacity, Melofolia, D-Day Land). Apparently heterogeneous, these fields of observation have the common factor that they are part of the same paradigm of local territorial development, based on the concepts of innovation, poles of economic attractiveness and territorial distinction. The cultural and creative dimensions of these projects appear in various forms: for some, they constitute a central argument in their promoters' discourses; for others, they are primarily a matter of symbolic capital mobilized in the discursive framework to justify the "innovative" and "creative" characteristics of these projects. Thus, the proximity of these projects to the cultural and creative industries and their local stakeholders is only of value here through their promoters' discourse. This choice is based on the notion of a "societal creative paradigm" (Moeglin 2019) which posits the centrality of certain specificities of the cultural and creative industries ("creativity", "experience", distinction, organization by "projects") in the communication and organizational practices of economic actors in all sectors.

Our methodology connects the sociosemiotic analysis of a corpus of more than 300 press articles[2] and institutional communications (websites, press releases, studies and reports) relating to these local projects, and approximately 30 interviews carried out with representatives of the main networks of local actors concerned by these projects. In addition to identifying the values that shape their discourses (Boltanski and Thévenot 1991), this corpus allows the production of maps of actor networks and schematizations, at several levels, of the power relations between these actors.

3.2. Local tensions generated by challenging the frames of thought and modalities of construction of public decision-making

Our main observations show a common context of a frame of thought about public action which, beyond each contested project, generates a mobilization of local actors who consider the decision taken to be incompatible with the socio-environmental issues faced by our societies. Three main common characteristics may thus be identified:

– These development projects and their contestation are observed in local territories experiencing industrial transition[3]. "Innovation", particularly focused on the digital economy and industries, is promoted as an essential lever for developing

2 Corpus produced via the Europresse database, from January 2018 to June 2021, among the local and national general and economic press.

3 Industrial transition is supported by national and international institutions, including the OECD, and aims to increase growth and competitiveness in regions that have suffered forms of deindustrialization. See: https://www.oecd.org/cfe/Regions-industrial-transition-highlights-FR.pdf.

the attractiveness of these territories on the grounds that it entails a knowledge economy, presented as localized, and would enable dynamics of growth.

– These territories and their decision-makers are at the same time subject to directions from national and international institutions to operate and support an energy and environmental "transition". The issues making these transitions necessary are of international scale. Hypermodernist neoliberal ideology, which constitutes the dominant frame of action in our capitalist economies, leads to a dissociation of local decisions from their global effects. In fact, the practices of local public decision-makers that we observed, conceived around objectives of growth and distinction, do not or only marginally integrate the effects of these ways of conceiving territorial development in terms of sustainable development, environmental protection or the fight against social inequalities.

– Beyond their particularities, the conflicts observed thus relate to local methods of managing management of this paradox, via power relations, between the search for "industrial innovation" in a capitalist economy and the urgency of a change in practices more favorable to the climate, the environment and the reduction of socio-economic inequalities (see Figure 3.1).

Figure 3.1. *Some forms of expression of the demands of opponent collectives in France – Imagiland Projects, Melofolia, Alibaba Liège and Belin-Beliet (source: ImagiNon collective, Chauffaille Autre collective, Stop Alibaba&Co collective). For a color version of this figure, see www.iste.co.uk/catellani/transitions.zip*

We propose below an analysis of the forms of mobilization in two types of public arenas (Badouard et al. 2016) within which the different stakeholders, placed in a situation of interdependence due to these local projects, articulate their discourses and their conflicts develop. The first of these arenas is media-related; the second is made up of public consultation mechanisms and bodies. These two forms of local public spaces allow us to better understand the forms of mobilization of these actors and to characterize the ideological registers of their respective positions.

3.3. Objects of the conflict configured by mediated discourses

In most of the cases observed, the press, especially local, plays a central role in both publicization of the project by its promoters, including public decision-makers, and in raising the profile (Voirol 2005) of the conflict that it generates. Sometimes relayed by the national press, the local media thus ensure that the economic prospects promised by the implementation of this project, and then secondly the conflicting issues mobilized by local actors in their discourse, are put on the agenda (Neveu 2010).

A typical media timeline may thus be characterized:

– A press release or a speech by a public decision-maker is relayed by the local media. Systematically, the arguments of job creation, the emergence or strengthening of economic or industrial sectors (e-commerce and logistics, tourism), and the expected gain in the territory's attractiveness are mobilized. This announcement is also, in some cases, relayed by the accounts of institutions or elected officials on digital social networks:

> For elected officials, this theme park is an opportunity for employment and the economy in the Saint-Yrieix region. There is no park based on this theme in the world, which is surprising. I have every confidence, it will be very innovative, this park will be focused on new technologies (Mayor Philippe Sudrat on the Mélofolia project, cited by France 3 Nouvelle-Aquitaine, December 20, 2020[4]).

– Civil society thus becomes aware of the project by way of this institutional and relatively formal announcement. Depending on the case, the degree of progress of the project (definition, contract with the public authorities, stages of construction, schedule) appears to vary. Certain development projects, the most successful, are thus presented as the result of a win/win partnership between an industrialist or investor and the public authorities. Others are first announced in terms of the land

4 See: https://france3-regions.francetvinfo.fr/nouvelle-aquitaine/haute-vienne/haute-vienna-musical-theme-park-project-melofolia-tient-cadence-1907496.html.

acquired or as rehabilitation projects for abandoned industrial sites, the very nature of the project, its financial arrangements and the implementation schedule remaining unspecified.

– This publicization of the project and the arguments mobilized to justify its relevance provokes, among certain residents, elected officials and activists, surprise (notably when the announcement comes after the public/private partnership has been sealed), concern (the project may generate potential risks) and mistrust (the project was not debated).

– The forms and rhythms of mobilization (from a few weeks to several months) differ greatly among the cases studied. However, a common process is observed whereby individuals and representatives of local organizations mobilize to collect more precise information about the project, the conditions for its definition and implementation. This demand is based in part on the mobilization of local media and digital social networks, in particular to challenge decision-makers. Very often, requests for clarification are countered with the argument of business secrecy.

– Gradually, local collectives (relatively small numbers, 10–50 people) form in order to bring together the information gathered on the project through various means. The lines of questioning or contestation which will configure the conflict thus emerge. The press generally relays the press releases of these groups in a balanced way, thus helping to establish power relations between the different parties as well as a framework for debate. Recurring reasons for these mobilizations relate to the artificialization of the soil or of the land acquired, the relevance of the project faced with local socio-economic issues, or a perceived "absurdity" in the face of international environmental issues.

– The project leaders develop, in various forms, strategies for avoiding conflict and denigrating opponents and, in the media as well as on social networks and press releases produced by the parties, a divide gradually forms between two paradigms of local economic development. Some prioritize dimensions of the economy, employment, innovation and competition between territories. The others oppose the impacts of the environment and on living conditions as well as symbolic dimensions of the project deemed "from another time", incompatible with major social issues, or disconnected from the economy and local actors. In certain cases, the general national media relay these oppositions, often in a factual manner. Often at the initiative of national or at least supralocal activist networks, and via digital tools, groups of opponents expand and develop their expertise, particularly around the technical and regulatory dimensions of the project:

> It is clear that the elected officials of our territory are still following schemas from the last century which are leading us to a dead end. Some elected officials need to open their eyes: the old world is finished. No offense to the General Council, the promoter and the

various stakeholders in this project, but we are no longer in the 1980s (Studios Occitanie, extract from the preliminary consultation report, CNDP, December 2020).

These different phases of mobilization of local actors faced with projects initially limited exclusively to the economic field contribute to broadening and diversifying the frame for analyzing their impacts. In addition to publicizing projects and relaying institutional discourse, local media to start with, then digital social networks later, thus contribute to producing conflict and making visible the divide between various modes of representation of the project and of relating to the world (Wolton 1992). This phenomenon, reinforced by the expertise of activist networks supporting local collectives, contributes to the politicization of local economic development.

This results, on the one hand, in the "delocalization" of the project which, originally perceived as anchored in living conditions and lived spaces, takes on meaning within a broader set of conceived spaces (Lefebvre 1974) where the privatized territory is envisaged as a resource for producing market value (Rosa 2020). The diversity of local mobilizations and their links at national and international levels thus enable the production of a common critical sense that goes beyond particularities to challenge a dominant framework of thought for local development. The forms of mobilization of opponents, especially the most conflictual and spectacular ones (demonstrations, petitions, occupations, blockades of sites), continue to mobilize the media, with the objective of winning public opinion and rallying it to its cause. The media thus contribute to changing the power relations between local actors.

3.4. Consultation mechanisms and bodies ineffective for shared construction of the project

French and European regulations provide for a set of consultation mechanisms to be set up as part of development projects. These mechanisms vary depending on the nature of the project, its land area or its budget, its geographical area of influence or interest, or even its environmental impacts. Different types of contextualized analyses are supposed to feed into these mechanisms (Blondiaux and Cardon 2006): impact studies, compliance with development plans and urban planning, compliance with administrative procedures. These bodies and mechanisms are perceived, and promoted, as constituting a regulatory arsenal guaranteeing, if not an open debate, at least scholarly and independent information on the effects induced by any large-scale development project. Although the effective organization of consultation mechanisms linked to any project is generally respected, their non-binding nature, their place in the project timeline, the types and quality of information actually made public and the effective methods of organizing the debate, mainly at the initiative of the project leader, only partially guarantee that the informed debate on ideas and facts will take place between local stakeholders (Charaudeau 2017). Finally, in their spirit and situated in their historicity, these different bodies have the objective as much of avoiding contestation as transparency of information for an open debate (p. 118). The frequent involvement of experts and the mobilization of "positive" terms aim, on the one hand, to make these projects acceptable, and on the other, to depoliticize them (Stein 2019) and neutralize their conflicting dimensions (Krieg-Planque 2010). Focusing on positive externalities by obscuring risks or negative effects, organizing the forms of spoken contributions so as to minimize controversies and conflicts, or imposing a timeline for debate aimed at its own avoidance (Carlino 2018) constitutes two recurrent normative characteristics observed during these moments that are presented as participative and adversarial.

In addition to these regulatory mechanisms, several other bodies are supposed to contribute to a shared and adversarial construction of projects. Firstly, representative bodies, at different territorial and administrative levels, are organized based on councils of municipal elected officials, at urban area or intermunicipality levels, and at supralocal levels. Less formally, initiatives to build counter-expertise by elected officials, residents or local associations and federations also allow debates to be initiated and legitimize requests for adaptation, or even rejection, of a project. These processes are long and generally require the support of extra-local actors specialized in certain dimensions of the project: environment, law, economy, etc. These forms of counter-expertise justify in themselves requests for debate, but appear to be little considered by the project leaders. On the other hand, and particularly regarding environmental law and administrative law, these steps make legal actions possible and admissible, for non-compliance with established procedures or for contradicting commitments at local, national or European levels.

Finally, forms of control over the territory, through temporary or prolonged occupation of sites, through street demonstrations and public awareness actions, complete the range of possibilities for consultation or co-construction of the project and constitute forms of "(counter)democracy" (Rosanvallon 2006). In the various forms mentioned above, power relations between the actors appear decisive. The communication actions of the different parties (via the press and digital media, via events, as well as via the resonance of the local project with others) then constitute the means of raising public awareness, clarifying information and, above all, making visible the project, the methods of its design and its potential impacts, in response to practices often masked under business secrecy or to strategies of "naturalizing" the decisions taken. These games of power, of power relations between local and extralocal actors, thus contribute to politicization of the project, whose process consists of reconfiguring its framework of perception and analysis. If the project's promoters present it by default exclusively in its economic dimensions (growth, employment, attractiveness of the territory), in themselves potential subjects of debate, the opponents develop "systemic" strategies, which link together all the social, environmental, financial, legal, political and strategic dimensions for the territory.

Our observations, again moving beyond the particularities of each case, highlight a set of practices, partly tacit, which tend to minimize the impact of these different forms of consultation and shared construction of the project:

– The contested projects observed are supported or initiated by public decision-makers (mayors, presidents of communities of municipalities, urban areas, regions) and rarely by the institutions that they represent. Personalization is strong and impacts the conditions of the debate.

– The representative bodies within these institutions are characterized by modes of organization, practices, even agendas, which do not allow, or even avoid, adversarial debate. Beyond elected officials' principled partisan positions, the relative "depoliticization" of these bodies reflects a certain verticality of decision-making as well as a relative ideological consensus on the technical requirements of growth, employment, commercial development of the territory and the disconnection of local decisions from major contemporary issues:

> The subject is taboo in the council. It is not a matter of discord, but it is somewhat unserious. The councilors are not openly keen or enthusiastic, but they bring some kind of discreet support, approval… but what I find pathetic about these stories is that apart from a few who have delved into detail, all the councilors relied on a kind of very *greenwashed* presentation, a one-man stand-up show by a representative of Vinci to make their decision (R. Manzanas, elected official from Angoulême, greater Angoulême urban area, about the Imagiland project)[5].

– The project promoters tend in their discourse, as well as in their representations of the territory and its development, to dissociate the local impacts of the project from its extralocal impacts (Harvey 2011). This ideological posture centered on direct economic impacts induces a form of waving away the consequences of decisions on the environment or structural social inequalities. Opponents try to exploit the mechanisms of consultation to break this rupture of levels:

> On the environmental aspects, as long as we have not solved the problem at European level, what is the point of being more Catholic than the Pope? If Alibaba didn't come here, they would move 100 km away from us, and environmentally speaking, not much would

5 Interview conducted March 6, 2021.

> change… except possibly the noise aspect (O. de Weisseige, Walloon Union of Companies, regarding the establishment of Alibaba at Liège airport)[6].

– The strategy of the project promoters, as acknowledged by certain elected officials and local economic actors, includes fine management of the project's timelines (Carlino 2018) and particularly its publicity: "Obviously, my interest is to speak as little and as late as possible to avoid giving fodder to people who are opposed. We are very discreet in our progress, but we are moving forward", says the promoter of one of the projects studied.

Thus, in all cases observed, the project made public is not only already conceived but also integrated within a territorial project and programmed in its implementation. The documents distributed combine both a partial and simplified presentation and technical elements intended to attest to the expertise mobilized. These materials, sometimes difficult to access, and the timeline of the forums for debate tend to limit the capacity of collectives to mobilize. The involvement of local and supralocal activist organizations, specialized particularly in environmental protection or administrative compliance procedures, allows rapid provision of expertise and construction of a critical analysis of the project justifying holding debates which are fundamentally adversarial.

These debates are supposed to be guaranteed by mechanisms of public consultation and dialogue. However, with the same objective of minimizing the capacity for counter-expertise, limiting spoken contributions, and reducing possible spaces for contestation, promoters of projects limit the periods for collecting opinions, and, in most cases observed, tend to avoid substantive debates. The number and frequency of consultation meetings, sometimes the choice of a limited number of places, a restricted time slot and a schedule leaving little time for discussion, make it possible to reconfigure these bodies for adversarial debate into moments of partial and unambiguous presentation of projects. In return, these practical conditions and strategies for circumventing adversarial debate themselves become sources of violent local conflicts; consultation mechanisms are then seen as bodies for registering projects that have already been sealed and are only amendable at the margins. Without having focused our study on projects which did not give rise to contestation, we were able to observe on the contrary that voluntary implementation of these consultation bodies, in advance of the construction of projects, constituted an element essential to the modalities of conflict management. For example, two years of real consultation with local residents and stakeholders have allowed consensual urban development in Miramas-Marseille. The establishment by the population of power relations for a constructive debate with

6 Interview conducted November 7, 2019.

public decision-makers has allowed the creation of cultural facilities in the Grand-Parc district of Bordeaux (Montero 2014).

3.5. Conclusion

We have offered here some analytical elements that may contribute to a better understanding of the conditions and forms of mobilization of local actors in opposition to economic territorial development projects, promoted by representatives of public institutions. Beyond their distinguishing features and their apparent heterogeneity, the projects observed all relate to representations of the territory as a market resource (for industrialists as well as for public decision-makers) and frames of thought for socio-economic development relating to a "creative paradigm".

Our research had a systemic ambition, to articulate the diversity of social and political phenomena, dynamics and processes that shape these decisions and conflicts. Here, we focused the presentation of the results of our observation concerning two dimensions of the manifestation of conflicts, two forms of oppositional public spaces: media coverage and consultation. We focused our discourse analysis on the forms of integrating environmental issues into these decisions and debates.

Media coverage of the projects, mainly provided by the local press and the stakeholders themselves, first allows their publicization, then the visibility of the conflicts they generate. For the various parties, the challenge lies in rallying public opinion. Although consultation bodies are sometimes set up, the effectiveness of these mechanisms generally does not satisfy opponents of these projects. These latter develop strategies, partly new and with the support of the expertise of extra-local organizations, with the aim of, on the one hand, curbing or stopping these local projects; on the other hand, reconfiguring frames of thought and representations, particularly by integrating environmental and social dimensions and by reconnecting local decisions to international issues.

Media coverage and the animation of conflicts within consultation bodies induce a phenomenon of politicization of projects, initially presented through rational arguments relating to their economic dimensions. The mobilization of opponents (residents, local associations, collectives, national NGOs), even though it often appears quantitatively weak, gradually tends to reconfigure the public conduct of the project, both through the advocacy of forms of conflict in opposition to consensual practices, and through its recontextualization in terms of contemporary social issues.

Thus, these bodies and moments tend to confront dominant representations and frameworks of thought for the construction of public decision-making with their limits and contradictions in the face of contemporary issues and aspirations for practices of consultation and collective production.

Finally, these mobilizations contribute to changing local and international power relations with public and institutional decision-makers. This trend aims in particular at the establishment of new methods of local governance, more collaborative and consultative, and proposing new statutes and methods of management of these territories, including public goods and commons (Ostrom et al. 2012; Dardot and Laval 2014).

3.6. References

Badouard, R., Mabi, C., Monnoyer-Smith, L. (2016). Le débat et ses arènes. *Questions de communication*, 30, 7–23. doi: 10.4000/questionsdecommunication.10700.

Blondiaux, L. and Cardon, D. (2006). Dispositifs participatifs. *Politix. Revue des sciences sociales du politique*, 75, 3–9. doi: 10.3917/pox.075.0003.

Boltanski, L. and Thévenot, L. (1991). *De la justification. Les économies de la grandeur*. Gallimard, Paris.

Bouquillion, P. (ed.) (2012). *Creative economy, creative industries : des notions à traduire*. Presses Universitaires de Vincennes, Saint-Denis.

Carlino, V. (2018). Temporalités de la controverse sur le nucléaire en Lorraine. Temps "long" des désaccords, temps "infini" des déchets radioactifs. *Questions de communication*, 34(2), 155–172. doi: 10.4000/questionsdecommunication.15722.

Cefaï, D. (2007). *Pourquoi se mobilise-t-on ? Les théories de l'action collective*. La Découverte, Paris.

Charaudeau, P. (2017). *Le débat public. Entre controverse et polémique. Enjeu de vérité, enjeu de pouvoir*. Lambert-Lucas, Paris.

Dardot, P. and Laval, C. (2014). *Commun. Essai sur la révolution au XXIe siècle*. La Découverte, Paris.

Harvey, D. (2011). *Le capitalisme contre le droit à la ville. Néolibéralisme, urbanisation, résistances*. Amsterdam, Paris.

Higgs, P., Cunningham, S., Bakhshi, H. (2008). Beyond the creative industries: Mapping the creative economy in the United Kingdom. Technical Report, NESTA, London.

Krieg-Planque, A. (2010). La formule "développement durable" : un opérateur de neutralisation de la conflictualité. *Langage et société*, 134(4), 5–29. doi: 10.3917/ls.134.0005.

Landry, C. (2000). *The Creative City a Toolkit for Urban Innovators*. Earthscan Publications, London.

Lefebvre, H. (1974). La production de l'espace. *L'Homme et la société*, 31/32, 15–32. doi: 10.3406/homso.1974.1855.

Mœglin, P. (2019). L'hypothèse d'un paradigme créatif sociétal. Intérêts, enjeux et stratégies en lice. *Communication*, 36. doi: 10.4000/communication.9702.

Montero, S. (2014). Concertation et tactiques citoyennes pour la création d'un équipement culturel à Bordeaux. *Participations*, 10(3), 151–176. doi: 10.3917/parti.010.0151.

Neveu, E. (2010). Médias et protestation collective. In *Penser les mouvements sociaux. Conflits sociaux et contestations dans les sociétés contemporaines*, Filleule, O., Agrikoliansky, É., Sommier, I. (eds). La Découverte, Paris, 245–264.

Ostrom, E., Chang, C., Pennington, M., Tarko, V. (2012). *The Future of the Commons: Beyond Market Failure and Government Regulations*. Institute of Economic Affairs, London.

Rosa, H. (2020). *Rendre le monde indisponible*. La Découverte, Paris.

Rosanvallon, P. (2006). *La contre-démocratie. La politique à l'âge de la défiance*. Le Seuil, Paris.

Stein, M. (2019). La controverse du "gaz de charbon" en France (2006-2018) : conflits de nomination et mise en question de la neutralité de l'expertise. *Mots. Les langages du politique*, 119, 69–85. doi: 10.4000/mots.24390.

Subra, P. (2016). *Géopolitique locale. Territoires, acteurs, conflits*. Armand Colin, Paris.

Vincent, A. and Wunderle, M. (2012). Les industries créatives. *Dossiers du CRISP*, 80(2), 11–90. doi: 10.3917/dscrisp.080.0011.

Voirol, O. (2005). Les luttes pour la visibilité. Esquisse d'une problématique. *Réseaux. Communication, technologie, société*, 129/130, 89–121.

Wolton, D. (1992). Les contradictions de l'espace public médiatisé. *Hermès, La Revue*, 10(1), 95–114. doi: 10.4267/2042/15362.

4

Mid-Mountains in Transition: From Tension to Relatedness

4.1. Mid-mountains in transition: context, questions and methods

4.1.1. *Transitions, deconfinements and links to the environment*

The reflective pathways and avenues of interpretation proposed in this chapter are fundamentally embedded in a context marked by singular characteristics. At the beginning of the decade, the mid-mountain regions underwent transformations linked to both situational elements (confinements/deconfinements, restrictions/modifications of activities) and significant trends. These include diversification of the activities practiced, the desire to reconnect with the natural environment (Chanvallon and Héas 2011), a strong orientation towards recreational aspects, the arrival of new audiences frequenting these leisure spaces, or further impetus from stakeholders in the resorts who embark in this path (Perrin-Malterre 2016).

At the same time, these mountain environments and territories, particularly attractive for recreational activities, are rich in remarkable, endemic and fragile biodiversity, undermined by climate change (Beniston 2003) and potentially by the renewal of human activities that take place there (Palomo 2017). These situational or tendential factors result from changes in the uses, representations and relationships of humans to their environment. Understanding these changes and the ability to

Chapter written by Amélie COULBAUT-LAZZARINI.

manage them represent a major challenge for regional stakeholders (Pailliart 2018), as evidenced by the open letter written by the mayor of Lans-en-Vercors[1].

In this context, stakeholders in the mid-mountain regions, through their environmental communication (Catellani et al. 2019; Coulbaut-Lazzarini and Couston 2021), highlighting their role as mediators, increasingly position themselves as upholders of a transition focused on conciliations and links.

4.1.2. *Transitions, questions and methods*

From a perspective of understanding these links, our thinking is based on a questioning that intertwines the consideration of the communication strategies implemented by actors in a mid-mountain territory with the way in which the communication thus developed aspires to transform society. This leads us to a series of questions. To what extent does communication on public issues such as the relationship between humanity and the natural environment in our contemporary societies succeed in de-centering public understanding and actualizing links? Do the modalities of this communication, particularly those which include sensory experiences, update ways of communicating, of making connections? More importantly, to what extent can these communication devices help build a relatedness-based model of society?

Through these questions, this chapter aims to understand how such communication may enable moving from a perception of environmental crisis, of tension, to the construction of a harmonious future, between linked individuals, groups of individuals and non-human actors. It includes consideration of intracorporeal (Martin-Juchat 2020), intercorporeal and interspecies communication.

From a methodological point of view, the results presented are based, on the one hand, on the analysis of communication documents produced or co-produced by the Bauges RNP (Regional Natural Park), and, on the other, on field surveys. These were carried out among “frequenters”[2] of the Bauges and Chartreuse (with the support of master’s students) and Vercors massifs, as well as various stakeholders (RNP, OFB (French Office of Biodiversity), guides, mountain herders). They particularly include the monitoring of human communication devices (eco-volunteers, citizen participation cycles), questionnaires, semi-directive interviews as

1 Open letter to the President of the Republic, the Prime Minister and the Prefect, 13 December 2020: https://www.facebook.com/michkrae/posts/10158066740307613.

2 “Frequenters” of the mid-mountain regions, according to our initial results, constitute a public that is mainly locals and/or over-educated and/or professionals in transmission or in mountain regions: a favorable breeding ground for promoting ideas and transformative actions.

well as designated experiences that were collected. The questionnaires were written collaboratively as part of the COMHUNENVI[3] project, based on the questionnaire developed by Léna Gruas for her thesis (Gruas 2021).

4.2. Mid-mountains, mediation and models of society: communication as basis for a common project?

Analysis of the actions and discourse undertaken by the regional natural parks (RNP) and the French Office of Biodiversity (OFB)[4] reveals that they are gradually developing communication and socio-environmental devices in a perspective of bringing the general public together around stories (Lits 2015) that update our models of society.

The official site of the Bauges massif RNP thus presents the park as a "territory of mediation". The project managers responsible for training the eco-volunteers who arrive during the summer describe it as a provider of "wonder for the visitor, peacefulness for those who live there" (training session, July 2021).

Through the elements of language collected in the RNP communication documents, a whole discourse constructs a passage from tension (linked to the multiple uses and challenges of the territory) to relatedness, envisaged as a triple relationship to ourselves, to others and to the world, which raises the question of links to otherness (Bolle de Bal 2003; Morin and Wolton 2011) and to the imagination (Coulbaut 2022).

From discourse to action, we observe the construction of messages and experimentation with communication devices, involving various actors, academic and socio-economic, from different disciplines or professional sectors, but who can converge around shared values (e.g. "The mountain… Respect!")[5]. We thus observe a desire to allow all frequenters of the territory to assume the values conveyed by these messages, by changing their representations and their actions. The link with the mountain environment and, more broadly, human belonging to the world, is also recalled, notably through explanation of the geopark label applied to the Bauges massif RNP. We thus learn that this geopark is "a state of mind that allows you to

---

3 *Communication et interactions humains-environnement* (human–environment communication and interactions) project in nearby Alpine natural spaces, co-financed by LABEX ITTEM, the Zone Atelier Alpes and the Ecology and Environment Research Federation.

4 See the new services created at the national level and dedicated to the development of communication devices to raise awareness among local authorities and decentralized services about human–natural environment links.

5 Slogan of the communication campaign begun in the early summer 2021.

gain a foothold within a very long history, that of the Earth. […] In a word, to better understand the long history uniting Man and the Earth"[6]. In this message, the place of humans can be read, one among others (Maris 2018) to which they are linked…

4.2.1. *Socio-environmental measures for implementing links*

In the Vercors RNP, the cycle of citizen participation on visitor numbers at Molière-Sornin (Vercors RNP/local actors) is an interesting example of a measure explicitly designed to implement links. Thus, during the cycle of citizen participation on visitor numbers at Molière-Sornin (spring 2021), the initial screening of a film featuring the voices of actors present in the territory (shepherd, hunter, climber, hiker, park manager, etc.) displays representations and uses in tension, for purposes of collective reflection on the activities that take place in this area. The debate and exchanges between actors at the end of the screening illustrate the efforts of the stakeholders to seek ways to reconcile their practices, not only with other human actors, but also with non-human living beings and the local ecosystem as a whole.

Among the communication systems designed and/or deployed by the Bauges massif RNP during the year 2021, we will briefly discuss three: two involving the summer period and one the winter period[7].

These examples illustrate the institution's tone and communication intentions, but above all reveal the structuring choices underpinning the dimension of mediation through this environmental communication.

4.2.2. *Humorous visual communication*

The first comes in the form of a communication campaign (Figure 4.1) co-developed[8] for the summer season, focused essentially on interactions with the herds of domestic animals present in the mountain pastures (e.g. understanding the importance of meadows for feeding herds, the role of fences, etc.).

6 See: https://www.parcdesbauges.com/fr/decouvrir-le-massif-des-bauges/s-emerveiller/l-esprit-geopark.html.

7 Since the first two have been the subject of a more in-depth analysis in previous works (Coulbaut 2022; Coulbaut-Lazzarini, forthcoming), we only recount the essential aspects for this chapter.

8 Campaign carried out by two RNPs, in collaboration with local actors and supported by the region and the European LEADER program.

Figure 4.1. *Example of a poster from the 2021 campaign in the Bauges. The poster plays on the fact that a sheep's plate of food is grass. The poster asks readers to not walk on the grass (or their plate of food). For a color version of this figure, see www.iste.co.uk/catellani/transitions.zip*

This was intended for the public who frequent these mountain pastures. The fundamental challenge of this awareness campaign was to prevent behaviors that generate tension and to replace them with behaviors promoting reconciliation of uses of space and respect for everyone, human or otherwise. The feedback collected from some mountain herders revealed a tangible expectation in terms of behavioral change in order to maintain or regain a harmonious cohabitation. Furthermore, adopting the perspective of animals, the campaign attempts a shift in point of view aimed at emphasizing the dimension of linkage ("Welcome to our home") rather than opposition or constraint. If this system may seem close to civic-tone campaigns that can be found in other contexts, it interfaces with the second, reinforcing the proposed visual message through human mediation.

4.2.3. *Human mediators to be found*

The second system deployed in summer is based on providing frequenters of the Bauges massif with "edutainment" activities led by environmental mediators and eco-volunteers. One of the peculiarities of this system is the absence of predictable location, allowing eco-volunteers to not reduce themselves to a kind of augmented signage, but to constitute an element of linking the audiences they encounter with their immediate environment. Like encounters with wildlife, meeting times with eco-volunteers and their educational kits remain fortuitous. This feature (also designed to avoid overcrowding of a particular site at a given time) invites hikers to explore and discover. Depending on their peregrinations, they may be able, perhaps, to pose questions to eco-volunteers, to use the texts or equipment made available to them on site (loan of binoculars at some sites) or to experiment with bird or soil recognition activities. Each time, the question of the impact of humans and the reconciliation of human presence alongside other living beings appears, whether centrally or more implicitly…

4.2.4. *From the visual to the sensory: guides, treasure hunts and giant animals*

Finally, the third system consists of a winter communication campaign (see, for example, Figure 4.2), centered on black grouse and chamois, co-developed with students of the master's program in Public Information-Communication and Media (ICPM). The latter, after having been able to interact with the director of a resort in the Bauges massif and four RNP project managers during meetings located on the mountain massif, designed different communication systems intended to be deployed over the coming winters. The campaign, which brings together different types of communication tools and systems, was commissioned to improve awareness among winter visitors of the emblematic fauna of the RNP. Through the choices made, we perceive a desire to immerse frequenters of the territory, through different types of engagement, in the environment within which they co-evolve with fauna with which they are largely unfamiliar. Visual elements such as the "little grouses' guide", puzzle-based treasure hunts or making chamois and black grouse more visible and perceptible through mascots resembling them, responsible for delivering messages and information about them, all perceptual and emotional dimensions are engaged to help visitors understand and feel the need to co-evolve with other species in the mid-mountain territories.

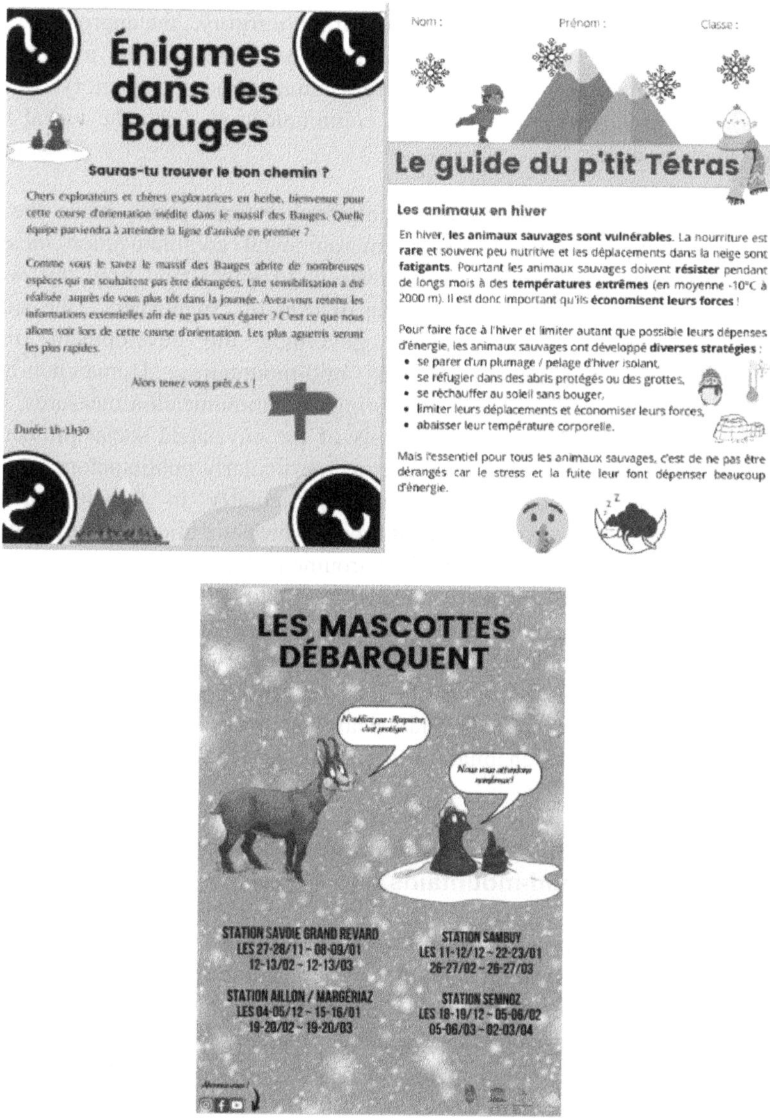

Figure 4.2. *Elements of the winter communication campaign of the Bauges RNP, proposed by students in the ICPM master's program. For a color version of this figure, see www.iste.co.uk/catellani/transitions.zip*

This campaign, like other measures, relies on the complementarity of communication tools and measures to constitute a mediation. By relying not only on visual hooks, such as can be conventionally observed elsewhere, but also, in parallel, on activities, integrating the lived experience of the territory, the approach is based on the physical and emotional involvement of frequenters of the mid-mountain regions and development of a link between humans and the territory. It is the interweaving of the different elements of communication, from the visual to the sensory, which mediates.

The programs studied, supported by regional natural parks, place the mid-mountain environment at the heart of environmental communication measures. Though future studies may seek more precisely to understand how this way of communicating relates to the understanding or uses of "frequenters", the analysis presented here sets out the elements governing the development of this environmental communication in the mid-mountains. Human/non-human interactions are involved in the very structuring of communication measures, where the ability to perceive otherness is the basis of the envisaged social project. The communicational strategies followed by actors, particularly public actors, on issues of the relationship to the living world (particularly in the mid-mountain environment) are based on explorations of types of measure allowing the creation of a link between the public and the natural environment. They are part of an attempt to define common horizons of expectation (Jauss 1978), to (re)create a story placing no longer the human, but human/non-human links at the heart of concerns, building transitions upon relatedness. The linking of all these actors at the heart of the mountain territory studied, as well as the link with the environment, are part of an attempt to modify representations and behaviors towards a complex ecological thought, into a "diplomatic cohabitation with the living world" (Morizot 2017).

4.3. Conclusion/questions for consideration: relatedness, key to development for the mid-mountains?

The programs and proposals led or coordinated by the Massif des Bauges RNP propose experimentation with diversified communication devices, chosen according to local contexts, the actors involved, but having the common point of the desire to act on representations and actions, in a transformative vision of society and of a relationship to the world. These focus not on tension, but on living together, on cooperation and, finally, on relatedness. They are anchored in experience, in Dewey's sense (Gégout 2014), for exploring links.

In mid-mountain territories, these measures reveal issues relating to coherence between representations and the reality experienced by program participants which are relatively close to those identified by I. Sourbès-Verger regarding space

(Sourbès-Verger 2002): the imagination finds a significant and different place depending on the actors involved.

The transitions taking place in the context of the mid-mountains raise questions about the interdependencies between actions and imagination. Questioning the imaginary then becomes a perspective for considering actions that may influence this imaginary, which, in turn, may generate new forms of action. Future research devices, currently being developed, will aim to deepen these perspectives.

To return to the path pursued in this chapter, it seems that the transition investigated here is a shift from a perspective of continuous growth in our Western societies to a perspective of balanced systemic co-evolution. We might think that the emphasis placed by institutions of the mid-mountains on relatedness and systemic co-evolution might hide a desire for euphemization or elimination of potential conflicts, thus acting as a rhetorical tool. However, this would be to misjudge these organizations, whose recent audit by the Regional Chamber of Accounts[9] confirms that certain parks have been able to effectively assert themselves as mediators in managing numerous conflicts of use on their territory (particularly mentioning the Bauges RNP concerning the preservation of species and agricultural activities). The study of the systems used by the actors mentioned situates environmental communication within a paradigmatic transition where relatedness, as a complex system of links, becomes the central element of societal dynamics.

4.4. References

Beniston, M. (2003). Climatic change in mountain regions: A review of possible impacts. *Climatic Change*, 59, 5–31. doi: 10.1023/A:1024458411589.

Bolle de Bal, M. (2003). Reliance, déliance, liance : émergence de trois notions sociologiques. *Sociétés*, 80(2), 99–131. doi: 10.3917/soc.080.0099.

Catellani, A., Pascual Espuny, C., Malibabo Lavu, P., Jalenques-Vigouroux, B. (2019). Les recherches en communication environnementale. État des lieux. *Communication*, 36(2). doi: 10.4000/communication.10559.

Chanvallon, S. and Héas, S. (2011). L'Homme et la nature : en quête/enquête sensible. *Natures sciences sociétés*, 19(4), 355–364.

Coulbaut, A. (2022). Coopération et reliance : design et communication environnementale comme logiques de pensée et d'action au service d'enjeux complexes. *Approches théoriques en information-communication (ATIC)*, 4(1), 127–137. doi: 10.3917/atic.004.0127.

9 See page 28 of the November 2022 topical public report: Rapport_public_thématique_CRC_ARA_les_parcs_naturels_régionaux_en_Auvergne_Rhône_Alpes_VF.pdf (ccomptes.fr).

Coulbaut-Lazzarini, A. (2024). Belonging to the world to think of the link: The relationship with the natural environment as a crucible of relatedness. In *Linking with Nature in the Digital Age*, Kohlmann, E. (ed.). ISTE Ltd, London, and John Wiley and Sons, New York, 191–208.

Coulbaut-Lazzarini, A. and Couston, F. (2021). Manifeste pour penser la communication environnementale à la lumière des liens humains/non humains. *Revue française des sciences de l'information et de la communication*, 21. doi: 10.4000/rfsic.10021.

Gégout, P. (2014). John Dewey (2013). Expérience et Nature. Paris : Gallimard. *Recherches & éducations*, 10, 267–268. doi: 10.4000/rechercheseducations.1978.

Gruas, L. (2021). Côtoyer les sommets, coexister avec l'animal sauvage. Contribution à la sociologie des pratiques sportives en milieu naturel. PhD Thesis, Université Savoie Mont Blanc, Chambéry [Online]. Available at: https://hal.archives-ouvertes.fr/tel-03544466.

Jauss, H.R. (1978). *Pour une esthétique de la réception*. Gallimard, Paris.

Lits, M. (2015). *Du récit au récit médiatique*. De Boeck Supérieur, Brussels.

Maris, V. (2018). *La part sauvage du monde*. Le Seuil, Paris.

Martin-Juchat, F. (2020). *L'aventure du corps. La communication corporelle, une voie vers l'émancipation*. Presses Universitaires de Grenoble, Grenoble.

Morin, E. and Wolton, D. (2011). Grand entretien. *Hermès, La Revue*, 60(2), 241–254. doi: 10.3917/herm.060.0239.

Morizot, B. (2017). Nouvelles alliances avec la terre. Une cohabitation diplomatique avec le vivant. *Tracés. Revue de sciences humaines*, 33, 73–96. doi: 10.4000/traces.7001.

Pailliart, I. (2018). Des territoires à la territorialisation. *Études de communication*, 50(1), 147–160. doi: 10.4000/edc.7635.

Palomo, I. (2017). Climate change impacts on ecosystem services in high mountain areas: A literature review. *Mountain Research and Development*, 37(2), 179–187. doi: 10.1659/MRD-JOURNAL-D-16-00110.1.

Perrin-Malterre, C. (2016). Processus de diversification touristique autour des sports de nature dans une station de moyenne montagne. *Mondes du tourisme*, 11. doi: 10.4000/tourisme.1012.

Sourbès-Verger, I. (2002). La tête dans les étoiles, les pieds sur Terre. Des représentations à la réalité. *Hermès, La Revue*, 34(2), 9–18. doi: 10.4267/2042/14423.

5

The Mountains in Transition: Introducing Scientific Culture into Tourism Under the Influence of "All Skiing"

5.1. Introduction

Within mountain territories, the introduction of scientific culture into tourism is enjoying a certain success, as evidenced by the multiplication of projects claiming to do so. This is the case, for example, for astrotourism within (multiple) recognized protected natural areas in the mid-mountains (Charlier 2018), glacial tourism in the high mountains (Salim and Ravanel 2020) or even geotourism (Bussart and Reynard 2022). These different forms of tourism product, grouped under the term scientific tourism, make the mountains a laboratory space for scientific mediation and research (Vialette et al. 2021). They have the common feature of being presented as possible pathways for territorial transitions, simultaneously as a tool for diversifying mountain tourism and as a lever for ecological transition in the mountains (Chambru et al. 2022). Apparently consensual, these projects can, however, become controversial, bringing into question the forms of territorial governance, the accessibility of scientific culture and their eco-compatibility (Chambru and Claeys 2024).

This chapter deals with one of these contemporary scientific tourism projects: the creation of a museographic space within the Dévoluy massif, in the Hautes-Alpes department in France. At first glance, it appears to be a form of eco-compatible territorial innovation responding to a double observation of the exhaustion of the "white gold" model and the early appearance of signs of climate change in the mountains, particularly marked in the Southern Alps (Collectif 2018). Based on a

Chapter written by Mikaël CHAMBRU, Cécilia CLAEYS and Quentin DAVEAU.

field survey carried out in 2019 and 2020[1], this chapter shows how this scientific tourism project shifts from an intention of diversification, a possible lever for revitalization of the historic village, to a strengthening of tourism centered on the alpine ski resort.

This project is part of a unique context, in terms of local heritage and contemporary politics (Achin 2015). Established in 1966, SuperDévoluy embodies the archetype of the so-called "integrated" ski resort advocated by the "snow plan". Based on the injection of large capital, technical efficiency and the massification of attendance, these modernist stations are established ex nihilo away from the historic villages. About 10 years later, a second station was created in a neo-mountain architectural style, but just as far from the historic villages – La Joue-du-Loup.

To this historical legacy of the "snow plan", diversification policies have today been added. From 2016, Hautes-Alpes has presented itself a pilot department for scientific culture in the mountains. For local elected officials, the stated ambition is to "succeed in scientific citizenship and tourism diversification". The Dévoluy museographic space – coupled with the NOEMA project[2] – is one of its flagship projects.

5.2. A tourism diversification project deprived of its scientific territorialities

At an altitude of 2,500 m, a network of radio telescopes managed by Institut de Radioastronomie Millimetrique (IRAM) has been built on the Bure plateau since 1985. This observatory has established itself as a focal point for astronomy among the world scientific community, and also as an iconic place for Dévoluy. Despite its relatively recent establishment, it is a territorial symbol for the inhabitants as well as for tourist communication media. It is around this highly symbolic scientific instrument that the mayor of the municipality of Dévoluy, also president of the community of municipalities of Buëch/Dévoluy, wished to create a museographic space in the center of the village of Saint-Étienne-en-Dévoluy. Intriguing in appearance, the Bure observatory is marked by a triple isolation from the territory upon which it is located. It is in the first place marked by geographical isolation

1 The research project "Mountain, communication and astronomy: public issues around science and tourism" (MAESTRO) funded by Labex ITEM as part of the "investments for the future" program managed by the National Research Agency (ANR) enabled 12 semi-structured interviews, associated with direct observation campaigns. Research report: https://maestro.hypotheses.org.

2 NOEMA (*NORthern Extended Millimeter Array*) is a Franco-German-Spanish research project aimed at developing the most powerful and sensitive telescope in the northern hemisphere, notably allowing the study of the birth of stars just after the Big Bang.

based on its limited accessibility to the public[3]. It is also marked by cultural isolation. IRAM has, in fact, chosen to give its scientific activities, which are very complex, little or no visibility. The geographical isolation of the Bure plateau thus appears as a metaphor for inaccessible knowledge. This inaccessibility is reinforced by a succession of fatal accidents linked to cable car access to the site[4]. This scientific isolation associated with these tragedies, still very present in local memory, may partly explain the abandonment of an initial scientific tourism project, imagined in 2002 in the context of reconstruction of a new cable car, envisaging the development of public reception facilities near the IRAM observation facilities.

In this context, the new project to set up a museographic space dedicated to astronomy in the valley and in the village center could have been part of a symbolic and spatial logic of providing secure accessibility to scientific culture. In addition, following an opportunity study carried out in 2013, the museographic space project is expanding, adding environmental sciences to the sciences of the universe. Indeed, Le Dévoluy features a dynamic Natura 2000 zone[5]. However, its scientific activity is barely visible locally and little recognized as such by the public, whoever they may be. This invisibility is explained notably by the fact that its scientific activity does not correspond to the representations these audiences have of what science is, in contrast to the activities of IRAM. It is also explained by the fact that the Natura 2000 zone is perceived by certain local actors as a measure external to the territory aimed at taking over management from its inhabitants in the name of a European policy of preserving ecosystems. Added to this is the confusion between scientific ecology and political ecology that resurfaces during local controversies[6]. These different elements explain why the Natura 2000 zone was absent from the original museum space project.

In 2014, the development of the scientific tourism project "Le Dévoluy from the underground to the stars" was presented as a tool allowing the territory to diversify its tourism offering. In this process of promoting tourism, scientific culture finds

3 It is impossible to access by car and only three trails along ridge lines and rockslides allow public access to it. A cable car serves the observatory, but it is private and the public cannot use it.

4 On July 1, 1999, a cabin detached and killed the 21 people inside. The disaster is one of the most serious involving a ski lift to have happened in France. In December 1999, the helicopter responsible for supplying the observatory got caught in the remaining cable of the cable car and crashed, leading to the death of its five occupants. In 2015, a new cable car was inaugurated, which again fell victim to an accident in 2016, without causing casualties this time.

5 Le Dévoluy is home to several rare, even endemic species, such as Bigot's weevil, a beetle found only on the Bure plateau.

6 The Dévoluy car rally, controversial for its environmental impact, is, for example, regularly an arena for this distrust of ecological activism.

itself mobilized as a public policy participating in territorial and economic development (Venzal-Barde 2006) before being a measure for making visible the cultural dimension of scientific activity (Urbas 2022).

5.3. A scientific tourism project derailed by the influence of winter tourism

The progress of this project had to deal with scientific actors who were reluctant to engage in scientific tourism as well as the non-existence of a previously constituted public of scientific tourists. This project also came up against a very strong local attachment to the epic story of "the-resort-that-saves-the-mountain" (Bourdeau 2021). In Le Dévoluy, this story is particularly embodied by a local figure, a farmer who became a ski instructor, a member of the municipal council on several occasions, and author of a book with an eloquent title: *SuperDévoluy l'épopée d'une invention collective (SuperDévoluy, the epic of a collective invention)*. And in fact, the ski tourism economy has clearly radically transformed the Dévoluy massif and continues to structure it today with 89% of second homes, 410,587 skier days in 2019, 582,931 winter tourist nights compared with 266,345 summer nights[7]. In this context, priority has been given by local elected officials to other tourist developments requiring strong economic investment, such as the creation of a balneotherapy center and the extension of the artificial snow network. The ambition is to continue to make Dévoluy an attractive tourist destination and to always welcome *at least* the same number of tourists. When it comes to the museographic space project, promotion of the scientific cultural heritage of the territory is understood as a link in an overall tourism offer that must be expanded to accompany its gradual diversification, while remaining articulated around the "all skiing" model (Bourdeau 2009).

Thus, in Le Dévoluy, scientific tourism is ultimately not understood as the possibility of implementing an alternative tourism system and even less a "post-tourism" allowing reinvention of a territorialized tourism, grappling with societal issues of the 21st century around "mountain-ness" (Bourlon et al. 2017). The "Dévoluy 2030" approach[8] aimed at formulating a strategy for the development of the territory by 2030 shows this very well: only 4% of the proposals made concern scientific tourism and only 5% concern fields other than tourism. After the municipal elections of March 2020, the project to establish a museographic space in Saint-Étienne-en-Dévoluy was completely transformed by the new municipal team. The museographic space has now been reduced to a minimalist version and moved

---

7 According to a study commissioned by the southern region on the Dévoluy skiing area.

8 Residents were invited to share their findings, ideas and projects through an online platform. A total of 94 proposals were submitted.

to the middle station of a new cable car linking the resorts of SuperDévoluy and La Joue-du-Loup at the top of the ski area, at 1,850 m above sea level, where the establishment of a new snow front is expected. This new equipment would be accompanied by the creation of 500 to 1,100 additional beds, distributed between the two stations. Scientific tourism then becomes the pretext for a policy of territorial development that tries to reconcile current modes of economic development with the climate emergency, rather than promoting alternatives to the white gold model responding to the socio-environmental challenges that the mountain region faces. Thus, in Le Dévoluy, tourist diversification in the mountains, even under the guise of scientific culture, is not necessarily a transition and even less an ecological transition (Chambru et al. 2022).

5.4. Conclusion

Academic literature has for several decades highlighted the economic limits of the white gold model and warned of the effects of climate change that have appeared particularly early in mountain areas (Bourdeau 2009). In response, after a phase of climate denial, adaptation strategies for mountain tourism have emerged, particularly in the Southern Alps, which have been more rapidly affected than the Northern Alps by the effects of climate change (Collectif 2018; Bonnemains 2019). However, these projects struggle to break free from the white gold model. Thus, the example of Dévoluy reveals how an alternative project to "all-skiing", aiming to "succeed in scientific citizenship and tourist diversification", was ultimately redesigned and put at the service of maintaining, if not strengthening, the ski resort model. It is an ultimate paradox that such strategies of adaptation to climate change through upholding the model tend to be supported by polycephalous public policies which finance at the same time both a climate plan aimed at reducing CO_2 emissions and a mountain plan encouraging the development of artificial snow, which is particularly energy-intensive. Citizen challenges emerge, however, which here and there manage to stem the headlong rush of the search for technical solutions. Such is the recent case of opponents of the project of a space of scientific mediation sheltering a hotel complex and its cable car on the Pic de Chateau-Renard located in the commune of Saint-Véran (Queyras, Hautes-Alpes) (Chambru and Claeys 2024).

Finally, the trajectory of the scientific tourism project in Le Dévoluy reveals the difficulties in integrating the diversification of mountain tourism within a dynamic of transition associating the greening of lifestyles with principles of social justice and knowledge sharing. Despite the very different configurations in which mountain territories find themselves, none of them escape these difficulties and the multiple demands for change which they entail (Bourdeau 2021). Those involved in mountain tourism, whether cultural, scientific or sporting, are increasingly taking up the imperatives for transition which are addressed to them. Some formulate projects reflecting their convictions and their capacities for action, ranging from alternative solidarity micro-projects to neoliberal greenwashing. Others get bogged down in forms of non-decision, as illustrated by the debates relating to the closure of the oldest ski resort in the Hautes-Alpes, Céüze 2000, a few kilometers from Dévoluy (Bonnemains and Claeys 2022). These controversies question the transitional capacities of a territorial development model, still dependent from a socio-economic point of view on winter tourist activity and where communication may be mobilized to avoid debate (Chambru and De Oliveira 2021).

5.5. References

Achin, C. (2015). La gouvernance de la diversification comme enjeu de l'adaptation des stations de moyenne montagne : l'analyse des stations de la Bresse, du Dévoluy et du Sancy. PhD Thesis, Université Grenoble Alpes, Grenoble.

Bonnemains, A. (2019). Stations de sports d'hiver et adaptation au changement climatique : une comparaison entre les Alpes françaises et suisses. In *Tourisme(s) et adaptation(s)*, Cholat, F., Gwiazdzinski, L., Tritz, C., Tuppen, J. (eds). Elya, Paris, 227–234.

Bonnemains, A. and Claeys, C. (2022). Saisir les opportunités de ses vulnérabilités ? Trajectoire territoriale d'une station de moyenne montagne, Céüze 2000. *Revue de géographie alpine*, 110–114.

Bourdeau, P. (2009). De l'après-ski à l'après-tourisme, une figure de transition pour les Alpes ? Réflexions à partir du cas français. *Revue de géographie alpine*, 97(3). doi: 10.4000/rga.1049.

Bourdeau, P. (2021). Dilemmes de transition. Les destinations françaises de sports d'hiver entre agir créatif, inerties et maladaptation. *Géocarrefour*, 95(2). doi: 10.4000/geocarrefour.18943.

Bourlon, F., Bourdeau, P., Michel, F., Inostroza, G. (2017). Le tourisme scientifique, un après-tourisme en Patagonie ? *Études caribéennes*, 37–38. doi: 10.4000/etudescaribeennes.11169.

Bussard, J. and Reynard, E. (2022). Géotourisme et médiation scientifique dans l'offre touristique en montagne. Le cas du bien UNESCO Alpes suisses Jungfrau-Aletsch. *Revue de géographie alpine*, 110–111. doi: 10.4000/rga.10269.

Chambru, M. and Claeys, C. (2024). Scientific tourism in multi-labeled protected areas: The ecological transition and controversy in the mountains. In *Biosphere Reserves and Sustainable Development Goals 1: Scientific and Practical Educational Issues in the Mediterranean*, Barthes, A., Cibien, C., Romagny, B. (eds). ISTE Ltd, London, and John Wiley and Sons, New York.

Chambru, M. and De Oliveria, J.-P. (2021). Changement climatique et développement territorial en montagne : conflictualités et enjeux communicationnels. Le cas d'une station de moyenne montagne des Alpes du Nord. *Revue française des sciences de l'information et de la communication*, 21. doi: 10.4000/rfsic.10306.

Chambru, M., Claeys, C., Lewis, N. (2022). La mise en tourisme de la culture scientifique en montagne : défis et enjeux pour des territoires en transition. *Revue de géographie alpine*, 110–111. doi: 10.4000/rga.10299.

Charlier, B. (2018). Vous connaissez les Pyrénées le jour, découvrez-les la nuit. Réflexions autour de l'artificialisation in visu des paysages célestes nocturnes pyrénéens. *Revue de géographie alpine*, 106(1). doi: 10.4000/rga.3874.

Collectif (2018). Impacts du changement climatique et transition(s) dans les Alpes du Sud. Working Group Booklet, GREC-SUD/AIR.

Salim, E. and Ravanel, L. (2020). Last chance to see the ice: Visitor motivation at Montenvers-Mer-de-Glace, French Alps. *Tourism geographies*, 25(1), 72–94.

Urbas, B. (2022). Culture scientifique. *Publictionnaire. Dictionnaire encyclopédique et critique des publics* [Online]. Available at: http://publictionnaire.huma-num.fr/notice/culture-scientifique.

Venzal-Barde, C. (2006). Culture scientifique et tourisme : exemple du réseau des "musées du Verdon". *Méditerranée*, 107, 87–92. doi: 10.4000/mediterranee.468.

Vialette, Y., Mao, P., Bourlon, F. (2021). Le tourisme scientifique dans les Alpes françaises : un laboratoire pour la médiation scientifique et la recherche. *Revue de géographie alpine*, 109–2. doi: 10.4000/rga.9337.

6

Offshore Wind Power on Facebook: "Reframing" Media Coverage of the Energy Transition

The news media are full-fledged players in debate around a controversy within public space (Fraser and Valenta 2001): they can increase or decrease visibility of the arguments constructed and defended by the various actors participating in the public debate (Marec and Babou 2015). They are thus able to affect the trajectory of the controversy itself (Badouard and Mabi 2015b). This capacity for action of the media is a source of mistrust for opponents of the offshore wind farm in the Saint-Brieuc Bay (Côtes-d'Armor, France), the beginning of construction of which in spring 2021 revived controversy and ignited protest. This wind farm is part of the government's energy strategy, focused on the transition to renewable energies. Fishing professionals working in the bay, local residents and associations for the preservation of local ecosystems question the merits of this transition and reject the dominant media framing (Entman 1993) which they consider unfavorable to their cause.

The regional media, characterized by closeness to the territory and the communities for whom they provide media coverage (Ringoot and Rochard 2005; Noyer et al. 2013), are particularly targeted by critics. Opponents of the wind farm thus turned to Facebook. Official pages, personal profiles or public groups are frequented spaces for discussion on the platform. For opponents, the challenge is to produce and publicize narratives that reflect their own account of events (Le Cam and Domingo 2015).

Chapter written by Catherine QUIROGA CORTÉS.

However, "the fate of a controversy is never played out in a single arena" (Badouard and Mabi 2015a). This consideration invites us to take an interest in the logic of interaction (Chateauraynaud 2011) which may exist between arenas of public debate (Dodier 1999). In the case of Saint-Brieuc, an exploratory observation of the Facebook spaces frequented by opponents of the wind farm reveals the activation by these opponents of such a logic: content from various media arenas is circulated there.

We aim to study this logic of interaction in order to understand the place occupied by regional media discourse in construction of an oppositional framing to the energy transition embodied by the offshore wind farm. We thus hope to help shed light on the role played by local and regional media in controversies around energy transitions.

6.1. Methodological approach

The analytical approach comprised three stages. The first step consists of observation of oppositional spaces on Facebook, aiming to measure the extent of the phenomenon observed in a preliminary manner, namely the sharing of regional media content. The two discussion spaces most heavily frequented on Facebook by opponents of the wind farm in the Saint-Brieuc Bay will be analyzed: the public group "*Ensemble pour un littoral français sans éoliennes*" (Together for a French coastline without wind turbines – 7,800 users) and the community page "*Gardez les Caps*" (Keep the Capes – 1,700 followers).

In the second step, we seek to determine the themes addressed in each discourse that we studied. We begin by putting together three corpora of data collected between April 3, 2021, one month before the start of the project, and November 3, 2021, when work was paused. The first corpus corresponds to the discourse produced by titles of the regional press, called "media discourse" or MD. It is made up of 305 articles obtained through the *Europresse* platform (keywords "Saint Brieuc wind farm") over a sample of five regional press titles (print and web editions)[1] which cover the geographical area of the Saint-Brieuc Bay. The second corpus, "oppositional discourse" or OD, corresponds to the discourse mobilized by opponents in the spaces studied on Facebook. It is made up of texts from 1,638 Facebook posts collected through the CrowdTangle[2] interface. The third corpus,

1 *Ouest-France*, *Le Télégramme*, *Les Presses d'Armor*, *Le Penthièvre* and *Le Petit Bleu des Côtes-d'Armor* (via Actu.fr).

2 Tool owned and operated by Meta for collecting public data from the social network Facebook.

comprising the elements of regional media discourse shared by opponents in these same spaces, called "shared media discourse" (SMD), comes from the media content shared by opponents in the Facebook spaces studied. Only the 501 publications from the OD corpus which included the Internet link (URL) of a regional press article were retained.

Next, we considered the textual data associated with the metadata of these URLs, which the CrowdTangle interface retrieves and classifies into "link text" and "description" categories, corresponding respectively to the titles and leads or first lines of the shared articles (see Figure 6.1). Although this limited corpus does not allow in-depth analysis of the contents of the articles shared, it does allow identification of the themes present in them. We retained 289 texts.

Link Text	Description
Saint-Brieuc. They create a collective opposing the "expensive, useless and dangerous" wind farm.	About a hundred people were present at Plerin on 28 October 2021 in the Côtes-d'Armor department for the creation of a citizen collective opposing the wind power project in the Saint-Brieuc Bay.

Figure 6.1. *Sample of text data associated with the metadata of URLs shared on Facebook, via CrowdTangle*

Then, we perform a lexical classification of each corpus through a textometric analysis using the IRaMuTeQ software (Ratinaud 2014). More precisely, we carry out a Reinert-type descending hierarchical classification (DHC) (Reinert 2008). The software lemmatizes and segments the texts, then orders these segments into thematic classes based on their degree of semantic proximity. These classes are presented in the form of dendrograms which show a list of the forms or words that are statistically most representative of the classified segments. The branches of the dendrogram indicate the proximity of the classes to each other and suggest links between the themes. Contextualizing these forms by reading the segments containing them is necessary to obtain a more detailed understanding of the themes identified and to carry out qualitative characterization.

Our last step consists of a comparison of the three discourses. After having determined the themes addressed in all the corpora, we constructed a comparative analysis grid (see Table 6.2) which indicates, for each theme (x-axis) and by corpus (y-axis), the lexical classes relating to this theme and the percentage of segments classified within them. Thus, comparison is based on the representativeness of the theme in each discourse and the way in which it is approached.

6.2. Results

6.2.1. *Observation of Facebook use*

Our observations of Facebook spaces (see Table 6.1) confirm the recurrence of sharing URL links to regional press articles.

Type of post	*N*
Total Facebook posts	1,638
Posts with URL	691
URL pointing to general press site	501
URL pointing to regional press article	307

Table 6.1. *Observation results for the Facebook spaces studied*

6.2.2. *Characterization and comparison of discourses*

The DHC analyses performed on each corpus are presented in the form of dendrograms. These show for each corpus the thematic classes present (11 for MD in Figure 6.2, 13 for OD in Figure 6.3 and 13 for SMD in Figure 6.4)[3], the distribution by proximity of these classes (two subgroups of classes for MD and SMD, four subgroups for OD), the percentage of segments grouped by class (e.g. 10.8% of the segments in the MD corpus in class 6 in Figure 6.2) and the most representative forms of each one (e.g. "impact", "study" in class 6 of MD in Figure 6.2).

An initial reading of the above-mentioned elements of the dendrograms allows us to deduce the themes addressed in each class of each corpus. A reading of the segments of each class using the IRaMuTeQ software interface allows us to confirm or correct the interpretation.

Thus, transversally, we identify a total of 13 themes addressed. Characterization of the themes present in each corpus is presented in conjunction with description of the comparative grid of the three discourses (see Table 6.2).

3 Classes 11 of MD and 13 of SMD are discarded because they contain metadata associated with irrelevant web articles.

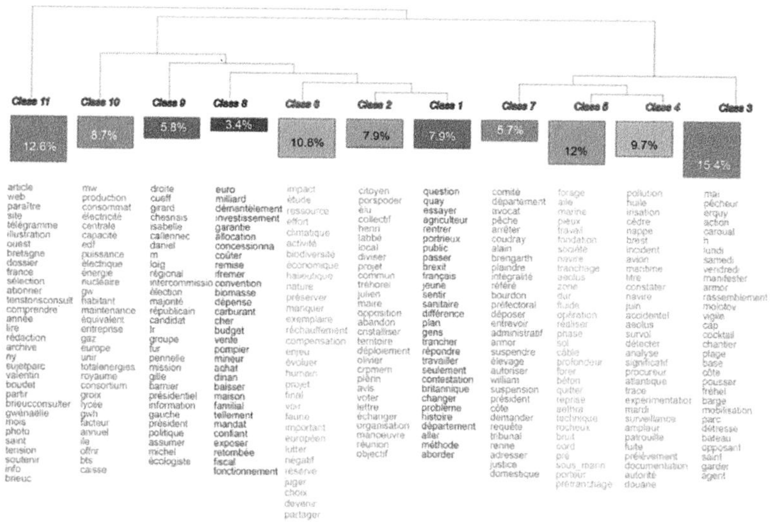

Figure 6.2. *DHC analysis of the MD corpus (IRaMuTeQ)*

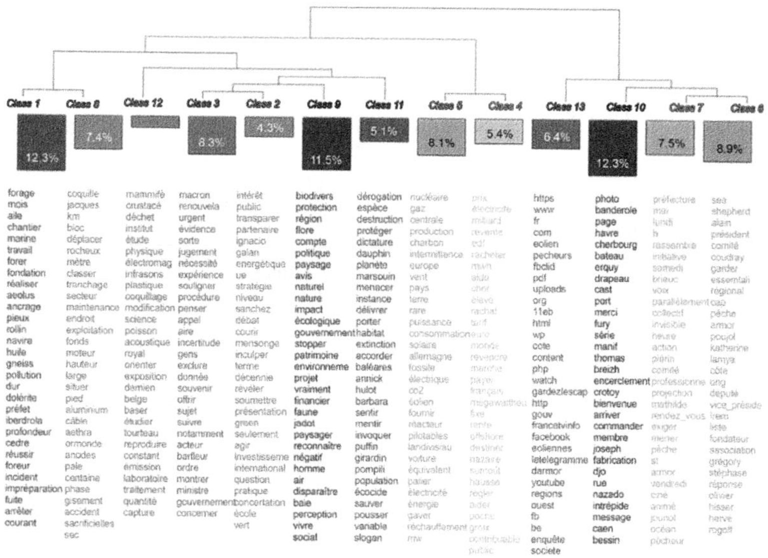

Figure 6.3. *DHC analysis of the OD corpus (IRaMuTeQ)*

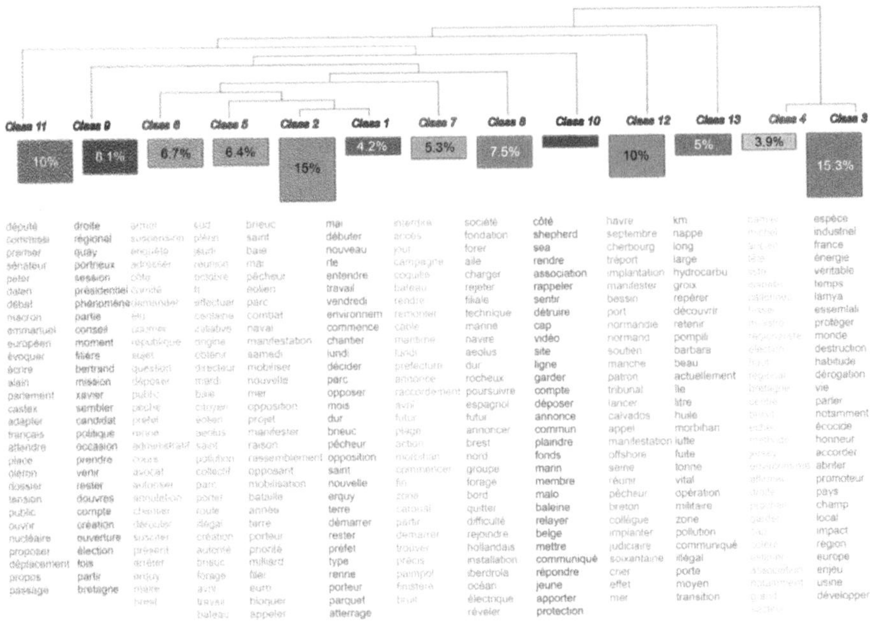

Figure 6.4. *DHC analysis of the SMD corpus (IRaMuTeQ)*

Category	Theme	MD (Figure 6.2)	OD (Figure 6.3)	SMD (Figure 6.4)
Energy	Production and consumption	**Class 10** 8.7%	**Class 5** 8.1%	–
	Economy	**Class 8** 3.6%	**Class 4** 5.4%	–
Politics	Regional	**Class 9** 5.8%	–	**Class 4** 3.9%
	National and European	–	–	**Class 11** 10%
	The Right	–	–	**Class 9** 8.1%
Ecology	Impact	**Class 6** 10.8%	**Classes 9, 11 and 12** 19.1%	**Class 3** 15.3%

Opposition	Figures	**Class 2** 7.9%	**Class 6** 8.9%	**Classes 5 and 10** 9.2%
	Repertoire of actions	**Classes 7 and 3** 21.4%	**Classes 3 and 7** 15.8%	**Classes 6 and 2** 21.7%
Works	Construction site	**Class 5** 12%	**Class 1** 12.3%	**Classes 1, 7 and 8** 17%
	Pollution	**Class 4** 9.7%	**Class 8** 7.4%	**Class 13** 5%
	Promoter	–	**Class 2** 4.3%	–
Other	Local issues	**Class 1** 7.9%	–	–
	Struggles	–	**Class 10** 12.3%	**Class 12** 10%

Table 6.2. *Comparison of corpora according to themes determined via DHC indicated are: the classes associated with the theme; percentage of segments grouped by class; "–" if no theme*

6.2.2.1. *Energy*

The theme of "production and consumption" of energy is present in the MD (8.7%) and DO (8.1%) corpora. Within MD, this theme is addressed through reported speech. The words of personalities from the political and socio-economic spheres, highlighting the advantages of the transition embodied by offshore wind, are confronted by those expressing skepticism. On the side of the opponents (OD), we find a purely critical discourse. Wind technology, compared to other energy sources, is considered of little relevance due to its low productivity and its ecological nuisance. Staying with these two corpora, we find the "economy" theme (MD 3.6% and OD 5.4%). MD discusses the details of the costs of construction and the electricity that would be produced. It also deals with fishing activity and the stock of scallops, a flagship species in the Saint-Brieuc Bay. In the OD corpus, opponents advance a new criticism: the cost of the energy produced by the wind farm is considered too high, as are the subsidies paid by the state to the developer. These two themes are absent from the SMD corpus. This seems to indicate that, by choosing not to share this part of the media discourse, opponents are blocking pro-wind farm discourse from access to the oppositional spaces they frequent on Facebook.

6.2.2.2. *Politics*

In the MD corpus, we find only mentions of the conflict between regional councilors around the wind farm. In the SMD corpus, we find mentions of regional political figures opposed to the wind farm, mainly situated on the right of the political spectrum. National (e.g. government members) and European political figures are also mentioned. No thematic class in the OD corpus refers to themes in this category. However, we find relevant lexical forms in diluted form in other classes. The political dimension is therefore not entirely absent from the oppositional discourse. This suggests that opponents rely on certain media content to challenge, criticize or express support for political figures. They thus manage to create a semblance of interaction with the political sphere.

6.2.2.3. *Ecology*

This theme is found in all the corpora. However, it is widely over-represented in the discourse circulating in Facebook spaces (OD, 19.1% and SMD, 15.3%). In the OD corpus, we find a very clear critique of the wind farm and its harm to the ecosystem of the bay and to endangered species. The articles in the SMD corpus seem to go in the same direction. Here, we find the words of different oppositional actors who denounce the "destruction" of marine ecosystems. This theme is less representative (10.8%) in the MD corpus, where it is addressed through the juxtaposition of two forms of speech: that of the supporters and that of the opponents. Thus, the harmful impact of the wind farm appears as an uncertainty around which debate is organized. These results seem to indicate that arguments contrary to the discourse of opposition are removed from the Facebook spaces. Opponents therefore do not seem to mobilize regional media content to support their arguments.

6.2.2.4. *Opposition*

The themes "opposition figures" and "repertoire of actions" are present in the three corpora in a more or less equivalent manner (nearly 8% for MD and between 15 and 21% for OD and SMD). This seems to indicate that opponents rely heavily on regional media content to promote their fight, which could suggest partial identification with the framing it provides. However, media content is filtered by the users of the Facebook spaces studied; regional press articles that report on actions which end in belligerent incidents are not reposted.

6.2.2.5. *Works*

"Construction site" occupies an equally significant place in the regional media discourse as in that of the opposition (MD 12% and OD 12.3%). It is very strongly represented in the SMD corpus (17%), which indicates a significant degree of centrality for news linked to the construction site in forming the oppositional argument. Pollution incidents occurring within the construction site are present in all corpora. The theme occupies a larger place in the MD corpus (9.7%) than in the SMD corpus (5%). For its part, the opposition discourse (OD corpus, 7.4%) points to the slowness of the work, as well as the developer's lack of preparation, accused of harming the local ecosystem. The theme "promoter" is exclusively present in this corpus (see class 2, Figure 6.3).

6.2.2.6. *Other*

Local issues not directly related to the wind farm (e.g. Brexit) appear only in the MD corpus, which indicates the centrality of the issues raised by the wind farm in the Facebook spaces studied. In these spaces, we also find mention of movements opposing other wind farms (OD 12.3% and SMD 10%). To support these mentions, it seems that content from regional media other than those in the sample is used. Opponents therefore seem to seek media content outside their information ecosystem to promote opposition to the energy transition represented by offshore wind farms.

6.3. Conclusion

We first confirmed the significant presence in Facebook spaces frequented by opponents of the Saint-Brieuc Bay offshore wind farm of content from the regional media field. This logic of interaction between media fields initiated by opponents is indicative of the centrality of the role played by regional media in publicizing protest speech and actions. These media appear, on the one hand, as legitimate sources of information, capable of allowing opponents to monitor the development of the wind farm and, on the other hand, as tools for legitimizing and promoting opposition to the wind farm. Finally, they also seem to act as intermediaries between two a priori distant spheres: the local oppositional sphere and the national political sphere.

By comparing the regional media discourse and that of opponents of the offshore wind farm in the Saint-Brieuc Bay on Facebook, we see significant differences in the way in which the energy transition is recounted. The most significant gap lies in the way in which the environmental impacts of the wind farm are approached: through a lens of ambiguity in the media studied and as certainties in oppositional discourse.

Furthermore, in something like a quest for media autonomy (Ferron 2016), the opponents appropriate journalistic logic for Facebook groups and pages, such as gatekeeping (Shoemaker and Riccio 2016) and media framing (Entman 1993). On the one hand, by sorting articles from regional media, opponents control access to the digital spaces they occupy. On the other hand, the selected articles are placed in the context of the protest and serve to support the opponents' arguments. If journalistic editorial production is a "reframing" of the experiences of the actors in the controversy (Esquenazi 2002), the oppositional discourse publicized on Facebook is in turn a "reframing" of the "mediatization" (Lafon 2019) of the wind farm.

6.4. References

Badouard, R. and Mabi, C. (2015a). Controverses et débat public : nouvelles perspectives de recherche. *Hermès, La Revue*, 73(3), 225–231. doi: 10.3917/herm.073.0225.

Badouard, R. and Mabi, C. (2015b). Le débat public à l'épreuve des controverses. *Hermès, La Revue*, 71(1), 145–151. doi: 10.3917/herm.071.0145.

Chateauraynaud, F. (2011). Sociologie argumentative et dynamique des controverses : l'exemple de l'argument climatique dans la relance de l'énergie nucléaire en Europe. *A contrario*, 16(2), 131–150. doi: 10.3917/aco.112.0131.

Dodier, N. (1999). L'espace public de la recherche médicale. Autour de l'affaire de la ciclosporine. *Réseaux. Communication – Technologie – Société*, 95(4), 107–154. doi: 10.3406/reso.1999.2157.

Entman, R.M. (1993). Framing: Toward clarification of a fractured paradigm. *Journal of Communication*, 43(4), 51–58. doi: 10.1111/j.1460-2466.1993.tb01304.x.

Esquenazi, J.-P. (2002). *L'écriture de l'actualité : pour une sociologie du discours médiatique*. Presses Universitaires de Grenoble, Grenoble.

Ferron, B. (2016). Professionnaliser les "médias alternatifs" ? Enjeux sociaux et politiques d'une mobilisation (1999-2006). *Savoir/Agir*, 38(4), 21–28. doi: 10.3917/sava.038.0021.

Fraser, N. and Valenta, M. (2001). Repenser la sphère publique : une contribution à la critique de la démocratie telle qu'elle existe réellement. *Hermès, La Revue*, 31(3), 125–156.

Lafon, B. (2019). Des médiatisations au processus de médiatisation. In *Médias et médiatisation. Analyser les médias imprimés, audiovisuels, numériques*, Lafon, B. (ed.). Presses Universitaires de Grenoble, Grenoble, 157–189 [Online]. Available at: https://www.cairn.info/medias-et-mediatisation--9782706142802-page-157.htm.

Le Cam, F. and Domingo, D. (2015). The plurality of journalistic identities in local controversies. In *Local Journalism: The Decline of Newspapers and the Rise of Digital Media*, Nielsen, R.K. (ed.). I.B. Tauris & Co. Ltd., London/New York, 99–116. doi: 10.5040/9780755695171.ch-004.

Marec, J.-L. and Babou, I. (2015). La dimension communicationnelle des controverses. *Hermès, La Revue*, 73(3), 111–121. doi: 10.3917/herm.073.0111.

Noyer, J., Paillart, I., Raoul, B. (2013). *Médias et territoires. L'espace public entre communication et imaginaire territorial*. Presses Universitaires Septentrion, Villeneuve-d'Ascq.

Ratinaud, P. (2014). IRAMUTEQ. Interface de R pour les Analyses Multidimensionnelles de Textes et de Questionnaires. Un logiciel libre construit avec des logiciels libres [Online]. Available at: http://www.iramuteq.org.

Reinert, M. (2008). Mondes lexicaux stabilisés et analyse statistique de discours. *Actes des 9e Journées internationales d'Analyse statistique des Données Textuelles (JADT). Lexicométrica* [Online]. Available at: http://lexicometrica.univ-paris3.fr/jadt/jadt2008/pdf/reinert.pdf.

Ringoot, R. and Rochard, Y. (2005). Proximité éditoriale : normes et usages des genres journalistiques. *Mots. Les langages du politique*, 77. doi: 10.4000/mots.162.

Shoemaker, P.J. and Riccio, J.R. (2016). Gatekeeping. *The International Encyclopedia of Political Communication*, Mazzoleni, G. (ed.). American Cancer Society, Hoboken, NJ, 1–5. doi: 10.1002/9781118541555.wbiepc202.

7

"*Y'a le feu au lac*!"[1]: Podcasts and Ecological Transitions

"The age of transitions" (Chabot 2015) involves not only the relationship of individuals to the planet, but also their place in society as well as the democratic role of the news media. Indeed, because the latter play an important role in "the co-constru[ction], entry into circulation and entry into debate of a set of constituted and available social discourses" (Delforce 1996, p. 27), they are clearly one of the predominant actors in the public space in which controversies about ecological transitions are intertwined. This chapter specifically aims to deal with the treatment of transitions through a booming media object – the podcast – and a particular approach: solutions journalism.

Even as public practices evolve, podcasting appears to be one of these new practices that journalists take on in the hope of rejuvenating their traditional relations with the public. According to Osmanian Molinero (2021), 15 million French people listen to podcasts every month. This format "has disrupted the production of the audiovisual content industries, by offering new media forms", according to Camille Brachet (2009, p. 21).

Certain themes stand out clearly in the landscape of these new productions, including the question of the energy, ecological and societal transitions. As "sequences of discussion and confrontation between divergent points of view on a subject" and as "a time of exploration and stabilization of issues during which the diversity of dimensions, the plurality of actors involved and possible courses of action are considered before political closure" (Lascoumes 2019), these media

Chapter written by Erika RIBERI and Pauline AMIEL.

1 "*Y'a le feu au lac*" is the title of one of the podcasts studied in this chapter. It plays on the French expression "*Il n'y a pas le feu au lac*" (the lake is not on fire), used to signify that there is no urgency in doing something.

objects participate in current controversies on ecological transitions (Krieg-Planque 2010). It is also possible to resituate the positions of Jean-Baptiste Comby in this paradigm of controversy as a sequence of discussions and conflicts between divergent points of view. These positions underline a trend towards individualization of public issues and depoliticization of issues related to energy and ecological transitions (Comby 2019).

On the contrary, solutions journalism presents itself as a possible and interesting way of approaching these topics for podcast producers. As a reminder, this is a professional approach based on principles of investigation and ethical rigor, which encourage journalists to break free of the traditional media framework of "bad news is good news" to propose potential solutions to societal challenges. This approach is therefore perfectly in line with the positive framework broadly defined by the term "transition". In addition, environmental issues as a theme are particularly addressed by solutions journalism (Amiel 2020).

The attraction of audiences to podcasts (Lombardo and Wolff 2020) and the approach of dealing with "solutions" thus make it possible to envisage this type of production as a relevant media object for trying to understand ecological transitions as a communication issue. In this chapter, we will therefore consider how podcasts on ecological transitions crystallize current issues in the communicational treatment of environmental issues.

7.1. Method

To answer this question, a corpus of native podcasts[2] dealing with ecological transitions with a "solution" positioning was first created. Podcasts that display a desire, in their presentation (paratexts or jingles, for example), to offer listeners solutions on questions of transitions were favored. To do this, a keyword search was first performed on the digital distribution platform Spotify. As this yielded a limited number of results, it was supplemented by a call for recommendations over the social-digital network Twitter. A total of 10 podcasts broadcast between 2017 and December 2021 were identified. They last between 15 and 80 minutes. Some include a limited number of episodes ("*Vivons heureux*", 7 episodes), while others include several hundred ("*Basilic*", 225 episodes).

2 "A native podcast is a podcast that was originally created independently of a broadcast radio program. It is 100% original" according to Serge Surpin in La Lettre A, accessed July 4, 2022, at 13:30. See: https://www.lalettre.pro/Le-podcast-natif-c-est-quoi_a16237.html.

Name	Origin		Creation date	Number of episodes	Duration	Status of presenter	Funding
Impacts positifs	Generalist media	*LCI*	2018	61	25–60 min	Sylvia Amicone, *LCI* journalist	Aldi partnership
Vivons heureux		*Arte radio*	2020	7	40–50 min	Delphine Saltel, journalist and producer	*Arte radio*
Y'a le feu au lac		*Ouest-France*	2020	21	10–20 min	Lucas Scaltritti, journalist and brand content manager	*Ouest-France*
Transition(s)	Media specializing in ecological transitions and sustainable development	*ID – L'Info Durable*	2019	21	10–40 min	Valère Correard, journalist director of *ID – L'Info Durable*	Partners in the business world
La voix des oasis		*Kaïzen*	2020	35	31 min	Alexandre Sattler, photographer	*Kaïzen* partnership, Colibris movement and Oasis cooperative

<table>
<tr><th>Name</th><th colspan="2">Origin</th><th>Creation date</th><th>Number of episodes</th><th>Duration</th><th>Status of presenter</th><th>Funding</th></tr>
<tr><td>Entreprises en mission</td><td rowspan="4">Media doubly specializing in economy/ entrepreneurship and sustainable development</td><td>Mediatico</td><td>2020</td><td>10</td><td>20–30 min</td><td>Frédéric Vuillod, business journalist and founder of Mediatico</td><td>Prophil consulting firm</td></tr>
<tr><td>On(ward) fashion</td><td>The Good Goods</td><td>2018</td><td>+ 100</td><td>15–90 min</td><td>Victoire Satto, co-founder of The Good Goods</td><td>The Good Goods</td></tr>
<tr><td>L'Empreinte</td><td>Badabam</td><td>2019</td><td>+ 100</td><td>20–40 min</td><td>Alice Vachet, expert in digital communication</td><td>ionos.fr partnership</td></tr>
<tr><td>Leur raison d'être</td><td>Ekopo</td><td>2020</td><td>44</td><td>15 min</td><td>Nathalie Croisé, journalist, conference presenter, trainer</td><td>Ekopo</td></tr>
<tr><td>Basilic</td><td>Independent podcast</td><td>Basilic</td><td>2017</td><td>225</td><td>30–40 min</td><td>Jeane Clesse, independent podcaster, founder of Basilic Lab, podcast creation agency, trainer</td><td>Basilic</td></tr>
</table>

Table 7.1. *Description of the podcasts in the corpus*

A qualitative analysis of the first episode of each podcast was then performed. For this purpose, a grid was created for comparing discourses and their construction (Ringoot 2014) and also their role in controversies over transitions. The objective was to consider not only the content of the podcasts, but also, based on the general principles of Mikhail Bakhtin's theory of dialogism[3] (Bakhtin 1984), "the effects of dialogism and echoes" (Riberi 2018) between podcasts and other discourses produced about transitions. They are thus analyzed in themselves and also within their overall ecosystem.

The analysis grid includes five criteria. The first allows us to analyze the narrative approach of podcasts (structure, editorialization and production of information). This is more specifically a question of identifying the journalistic format adopted (interview or report, for example) and also of identifying the potential presentation of journalists/podcasters.

The other four criteria are based on the criteria for solutions journalism as defined by Pauline Amiel in *Le Journalisme de solutions* (2020). This allows us both to look at the actual participation of the podcasts studied in the codes of this journalistic approach, and to observe, both in substance and in form, their treatment of the question of transitions. The following were studied: the provision by podcasts of critical information marked (1) by rigor and (2) by investigation; providing credible answers to societal problems, by presentation to listeners (1) of the subject, (2) of its effectiveness and (3) of its limits; the desire to increase public involvement; finally, the evident desire to play a role and have an impact in helping solve societal problems. For analysis of this last criterion, the two dimensions mentioned by Jean-Baptiste Comby in how ecological issues are dealt with, namely individual and political, served as a conceptual framework for our work (2019). The objective was to see to what extent media coverage reinforced the depoliticization effects presented by the researcher.

---

3 M. Bakhtin (1895–1975), whose work touches as much on the theory and history of literature as on linguistics, was particularly interested in the communicative function of language. In this sense, the utterance, which he considers to be "the real unit of verbal exchange" (Bakhtin 1984, p. 298), must be envisaged according to him as "a real link in the chain of verbal exchange, within a given sphere of human reality or daily life" (p. 290). In this way, an utterance always responds to previous utterances and may itself give rise to new responses, which will take the form of new utterances. For Bakhtin, however, an utterance is also "filled with echoes and reminders of other utterances" (p. 298): "it has clear boundaries, determined by the alternation of speaking subjects (speakers), but within the limits of these boundaries, the utterance […] reflects the verbal process, the utterances of others and, above all previous links (sometimes close to each other, but which may also be distant in areas of cultural exchange)" (p. 301). Understanding a statement is therefore impossible without considering the "dialogical harmonics" (p. 300) which fills it and passes through it. It is also important to note that, in Mr. Bakhtin's thinking, a novel or a scientific treatise is considered as utterances in the same way, for example, as a line of dialogue, presented as "the classic form of the verbal exchange" (p. 278).

7.2. Results

In order to explore in more detail what the podcasts in the corpus say about the communicational treatment of environmental issues, we use the perspective of Pierre Lascousmes' definition of controversies and, by extension, of the previously presented positions of Jean-Baptiste Comby. The aim is to see to what extent they make it possible to consider, as a "time for exploration and stabilization of the issues" linked to transitions, (1) "the diversity of dimensions", (2) "the plurality of actors involved" and (3) different "possible courses of action" (2019).

7.2.1. *Diversity of dimensions*

Firstly, the approach to the question of transitions through the individual and political dimensions as emphasized by Jean-Baptiste Comby appears clearly when analyzing the podcasts. However, this also reveals a third dimension which is strongly present, since it appears in different forms[4] in no less than eight podcasts out of ten: the economic dimension. In Comby's research upon which the analytical work is based, the political dimension in fact looks more specifically at public action carried out with a view to organizing the social world and is therefore distinguished strictly speaking from the economic dimension[5]. Between individualization and politicization, the economic ecosystem is thus established in the podcasts of the corpus as a major factor in mobilization around issues related to transitions.

7.2.2. *Plurality of actors*

A wide variety of people are interviewed in the podcasts, including researchers, journalists, entrepreneurs, members of associations or NGOs, citizens or communities. Within this plurality of stakeholders, however, we see some notable absences: political representatives. None appeared in or were invited onto the podcasts of the corpus.

4 For example, entrepreneurs may be interviewed to raise public awareness of transitions or mentions of the economic ecosystem can be made by different types of actors to take up actions neglected by politics.

5 In this perspective, the researcher says: "[the depoliticization of the social world] is manifested in the multiplication of public policies that address collective problems by managing individual conduct. These measures which incentivize or punish are the result of politicization (the political authorities taking charge of the problem) which depoliticizes (the problem requalified as not political but individual, technical, economic)" (Comby 2019). The researcher himself thus suggests this distinction between politics and economics. However, he does not develop it, in favor of an opposition between individualization and politicization of ecological issues.

From another perspective, in which the podcast is considered as a "sequence of discussion of divergent points of view" on a subject (Lascousmes 2019), we also note that there is little polyphony within the podcasts themselves. This is only present in two productions, "Y'a le feu au lac" and "Vivons heureux avant la fin du monde", which also present the same format: the journalists set the scene by asking themselves a question relating to transitions and thus move on to meet different stakeholders who might enable them to answer it.

The other podcasts all have the same format, namely the interview. While this format is inexpensive, it does not, by definition, allow several points of view to be heard on the subject, except by allusion or reference. Moreover, even the latter approach is rarely used in podcasts.

7.2.3. *Possible courses of action*

The courses of action considered in the podcasts will fall into one or more of the previously defined dimensions: individual, political or economic. Overall, and on an initial level, the podcasts seem to confirm Comby's analyses, since we observe a strong preponderance of individual courses of action: all the podcasts address it in their first episode, without exception. The political course of action is also the least present (in only 4 out of 10 podcasts). Furthermore, it is never approached, at least initially, as the main course of action.

By refining the analysis, four different types of courses of action which more or less combine the different dimensions may be distinguished:

(1) Firstly, it is possible to identify a so-called "inspiration" course of action: these are the podcasts which only rely on the individualizing dimension ("Basilic" and "Transitions").

(2) Next come podcasts combining the individualizing and the economic, which present the individual creation of products or businesses in favor (or on the pretext?) of transitions ("On(ward)fashion", "Impact positif", "Leur raison d'être" and "L'Empreinte"). These first two courses of action are all considered and dealt with by journalists in the interview format. It is also possible to note that none of these podcasts really meets, beyond the advertised posting, the different criteria for solutions journalism. Ultimately, they seem closer to what could be termed "inspiring positive journalism": in the discussion, the attention of journalists and questions is indeed directed more at the respondents than at the solutions they promote, almost promoting them as flagbearers and inspiring leaders of the transition. The teaser of "Impact positif" shows this well, being presented as:

> The podcast that tells you about these ordinary people who do extraordinary things. About those people who want to have a positive impact on society, those

who inspire, those who simply do. In recent years, a million revolutions have arisen, from the citizen, to the entrepreneur, including large groups or even students, the wave is felt everywhere: welcome to this world in transition, where everyone, at their own level, can move the boundaries. [...]

(3) A third type of course of action combines individualization with a more or less strong political coloring ("Vivons heureux avant la fin du monde", "Y'a le feu au lac", "La voix des oasis"): this one outlines the course of action for system change through degrowth and an exit from capitalism, even though this is not always assumed. Thus, for example, the journalist from "Y'a le feu au lac", while he addresses the question of digital pollution, only sketches the possibility of degrowth, without ever really naming it or arriving at it completely. Conversely, the first episode of "Vivons heureux avant la fin du monde", dedicated to fast-fashion, is much more explicit on the need to question neoliberalism and for intervention by politicians in this process, which is done through the intervention of the historian Jean-Baptiste Fressoz:

It is all very well, of course, to advocate a form of reduction in individual consumption [...]. The danger of this discourse is that it creates new markets for ethical clothing [...] and above all it gives the impression that it is only a consumer issue, and therefore an individual matter [...]. When in fact, it is mainly a matter of the mode of production and of acting upon industry very strongly [...], let's say to act through norms, through administration, through politics, through regulation, through rationing, why not.

(4) Finally, the podcast "Entreprises en mission" outlines a fourth course of action, which covers all the dimensions and synthesizes the various previous courses of action by presenting companies as a new place of general interest. The course of action envisaged involves the testimony of the guest Virginie Seghers, president and co-founder of the consulting firm Prophil[6]. Here, she affirms her own individual commitment to rebuilding the company as a "place for preserving the common goods of humanity"[7].

This re-foundation involves politics, since political action has allowed a reform of the Civil Code and the creation of a new status of company ("entreprise à mission"), established by French law in 2019, which Virginie Seghers presents in the podcast:

---

6 "A consulting firm that deciphers economic models in the service of the common good and supports entrepreneurs in their engagement" (Spotify).

7 Thus, she states: "In my professional life, I have always chosen to combine private initiative with the general interest. [...] My commitment has always been in fact to fulfil what we might call today a mission, which is to try to transform the dominant economic models, and therefore in this case the capitalist model, to make it more virtuous, more responsible, more sustainable".

The mission-driven company is a company, therefore, which is transformed in law to adopt a new legal form. In its social purpose, it is committed to pursuing a mission extended to social and societal issues, thus revisiting the notion of a company's social purpose. And moreover, it sets itself objectives, it makes commitments in order to respond to this mission and maintain it over time, and it engages its stakeholders in a process of co-construction and opposability of the mission.

However, it should also be noted that, in a certain way, this passage through policy and legislation is also intended to replace it: ultimately, the initiative to change the economic and political system comes from the company's voluntary approach and not from the will or decisions of political leaders. In this sense, this fourth course may also confirm a certain tendency towards the depoliticization of ecological issues.

7.3. Conclusion

In conclusion, the results revealed in this study allow advances in research on the communicative treatment of environmental problems in that they confirm, firstly, a strong tendency towards depoliticization and individualization of these issues. Above all, they make it possible to uncover a third dimension that is particularly fruitful for addressing possible courses of action for working towards environmental transitions: the economic dimension.

Analysis of the content of podcasts allows us to draw conclusions that are already very rich. However, as Hansen (2011) suggests and in order to better understand the communicational treatment of environmental issues, it would be appropriate to consider more broadly the study of podcasts from their production to their reception, which the limited length of this chapter has not made possible.

Indeed, by using the codes of solutions journalism, podcast producers also position themselves vis-à-vis traditional media to counter an approach to ecological transitions deemed anxiety-provoking, thereby questioning the creation of news information. Shaken by the emergence of new digital players and by new uses of information by the public, the news media are also experiencing a transitional phase. It would therefore be interesting to explore how the podcasts in the corpus, beyond the issues linked to the communicational treatment of environmental issues, also crystallize the current changes in journalism.

Finally, the difficulties in identification encountered during the creation of the corpus have already raised the question of the audiences and reception of podcasts. We should therefore also consider how they are received, and by whom, in order to really measure the impact on society of these media messages on environmental issues.

7.4. References

Amiel, P. (2020). *Le journalisme de solutions*. Presses Universitaires de Grenoble, Grenoble.

Bakhtine, M. (1984). *Esthétique de la création verbale*. Gallimard, Paris.

Brachet, C. (2009). L'appropriation d'Internet par les médias "non informatisés" : le cas des podcasts. *Communication langages*, 161(3), 21–32. doi: 10.4074/S0336150009003032 [Accessed 21 October 2022].

Chabot, P. (2015). *L'Âge des transitions*. Presses Universitaires de France, Paris. doi: 10.3917/puf.chabo.2015.01 [Accessed 21 October 2022].

Comby, J.-B. (2019). Retour sur la dépolitisation des enjeux écologiques. In *Manuel indocile de sciences sociales. Pour des savoirs résistants*, Copernic, F. (ed.). La Découverte, Paris, 470–480 [Online]. Available at: https://www.cairn.info/manuel-indocile-de-sciences-sociales--9782348045691-page-470.htm [Accessed 21 October 2022].

Delforce, B. (1996). La responsabilité sociale du journaliste : donner du sens. *Les Cahiers du journalisme*, 2, 16–33.

Hansen, A. (2011). Communication, media and environment: Towards reconnecting research on the production, content and social implications of environmental communication. *International Communication Gazette*, 73(1/2), 7–25. doi: 10.1177/1748048510386739 [Accessed 21 October 2022].

Krieg-Planque, A. (2010). La formule "développement durable" : un opérateur de neutralisation de la conflictualité. *Langage et société*, 134(4), 5–29. doi: 10.3917/ls.134.0005 [Accessed 21 October 2022].

Lascoumes, P. (2019). Controverse. In *Dictionnaire des politiques publiques*, 5th edition, Boussaguet, L., Jacquot, S., Ravinet, P. (eds). Presses de Sciences Po, Paris.

Lombardo, P. and Wolff, L. (2020). Cinquante ans de pratiques culturelles en France. *Culture études*, 2(2), 1–92. doi: 10.3917/cule.202.0001 [Accessed 21 October 2022].

Osmanian Molinero, L. (2021). Music, podcasts and proximity: The radio's key strengths. *Médiamétrie*, June [Online]. Available at: https://www.mediametrie.fr/fr/music-podcasts-and-proximity-radios-key-strengths [Accessed 24 June 2024].

Riberi, E. (2018). Barcelone, Marseille et Naples, territoires du picaro urbain : modéliser un discours littéraire de la marge. PhD Thesis, Aix-Marseille Université.

Ringoot, R. (2014). *Analyser le discours de presse*. Armand Colin, Paris.

8

Analyzing an Ecological Controversy on YouTube: Obstacles and Methodological Issues

8.1. Introduction

According to the collapsological thesis popularized in France by Pablo Servigne and Raphaël Stevens (2015), humanity is now irreversibly altering its environment, thus endangering the sustainability of nature's cycles and survival on Earth. This thesis, which is often found in ecological debates, is still the subject of constant controversy, leading certain researchers to investigate these moments when major societal issues are discussed: thus, the work of Francis Chateauraynaud (2011b) on the place of nuclear power in the climate debate and that of Lionel Scotto d'Apollonia (2014) with his socio-epistemic analysis of climate controversies. Among this type of controversy, there is increased interest in the problem of deforestation in the Amazon, not only because of the exceptional extent of this tropical forest and its place in the global ecosystem, but also the economic, social, political and geopolitical issues in which it is implicated (Dagicour 2020).

In information and communication sciences, beyond these ecological controversies, a lot of research focuses on the challenges of investigating them on

Chapter written by Bi Mathieu TRA, Jean-Claude DOMENGET and Orélie DESFRICHES-DORIA.

the Web, especially as this space has become, with the hegemony of social media[1], the top priority space for expression of social debates and because these debates constitute the ideal framework for observing the construction of social life. Tommaso Venturini's work (2012), which analyzes the unfolding of an online socio-technical controversy, concludes that trying to represent the complexity of social debates in a legible form remains a challenging endeavor, all the more since mapping methods are far from fully developed and far from achieving the goal of observing the whole complexity of controversies and making them easily legible. An alternative to the graph method has been proposed by Francis Chateauraynaud and David Chavalarias (2017) in their analysis of social media in the light of a pragmatic sociology of controversies. But instead of an alternative, we argue that the complementarity between these two methods cannot be sufficient, even though it appears to be a relevant solution for expressing controversies.

We therefore advance towards a third methodological pathway to constitute a triptych, made up of the two methodologies mentioned above, supplemented by a screen writings (*écrits d'écran*) approach. This latter, from a position of general theory of computerized media, allows us to investigate forms of publishing within the spaces engaged in the controversy and to consider a media totality representative of the actions and uses likely to influence the controversy. The objective of this chapter is to shed light on some of the obstacles and methodological issues that have guided the study of the ecological controversy on YouTube around forest fires in the Amazon. This proposed approach should allow better identification of the various forces present in an ecological controversy on socio-digital media. Above all, it proposes a solution to the problem of investigating controversies on YouTube by combining software tools with the methodological approach that we will outline, backed by the theory of screen writings proposed by Souchier et al. (2019).

After proposing an epistemological framework for analyzing a controversy, we will move on to testing on YouTube a formal model of exploration of the media actors in an ecological controversy with graph tools, on the one hand and, on the other hand, a socio-informatics approach to controversies with the Prospero software, while considering the media depth in the socio-technical dimension of the media and their uses.

---

1 Similarweb. Website Traffic Analysis [Online]. Available at: https://www.similarweb.com/corp/lps/similarweb-traffic-analysis/. Still in 2021, YouTube remains the leader in social media with 35 billion visits. The choice of YouTube is also due to the fact that through its comment sections, it offers a collective critical space where controversy over the Amazon forest fires unfolds, and because it is a form of social media where the power of images is expressed.

8.2. Addressing a controversy on socio-digital media: epistemological pluralism

It is customary to justify the use of multiple theories in the humanities and social sciences by several constraints, the first of which is that the objects here are complex, multidimensional and intertwined. But the other reason for epistemological proliferation specifically concerns information and communication sciences (ICS, SIC in French), due to their interdisciplinary character. ICS embrace several areas of research, each of which has its own research channels, paradigms and methods.

If epistemological theories account for the way in which knowledge can be understood, they also make it possible to measure the value of the knowledge produced in a given discipline as behind each theory there are ontological commitments, i.e. hypotheses about a vision of real-world objects and their perception.

What is a controversy and what are the most interesting epistemological approaches to judging an ecological controversy in a socio-technical space like YouTube? Juliette Rennes (2016, p. 28) defines controversy as "a polarized, argued, reiterated and public discursive confrontation". Starting from a base of criteria including status, theme and scientific, doctrinal and social positioning, Patrick Charaudeau (2015) establishes a typology of controversies. He distinguishes between a scientific and technical controversy, whose issue is an issue of truth, a doctrinal controversy, an ideological controversy and a social controversy which constitutes the most cross-cutting set of controversies. Social controversy, according to Charaudeau, takes place in a plural public space. It focuses on social events or on moral issues where the protagonists defend antagonistic positions with respect to the event, the subject of the controversy. For Rennes, a political controversy[2] has its source in the violation of a principle of justice or in non-respect of a general interest. Depending on the protagonists, this endangerment may consist of the preservation of a state of affairs or its change. Political controversy has cross-cutting issues because, according to her, many values may be defended or challenged here. Ecological controversies, such as forest fires in the Amazon, are today among the social controversies. Charaudeau points out that, in practice, certain social controversies may lead to scientific controversies, which in turn will lead to research in order to resolve the initial social question or that the relationship between social controversy and scientific controversy can lead to the ideologization of certain social behaviors.

Desfriches-Doria et al. (2018, p. 5) propose a synthetic definition of the previous approaches:

2 Charaudeau refers to social controversy.

> Controversy is considered as a social event: individuals and institutions take part in it, implying varying levels and natures of "concern" (Chateauraynaud 2007, 2011a). They deal with topics that involve knowledge with a sometimes high level of technicality, which may have an impact on choices regarding possible models of society. They are also characterized by exchanges of arguments, and varying conceptions of particular interest and general interest (Dewey 1927). The conditions in which the debate takes place influence these exchanges (Elster 2005), and consequently the nature of the power relationships involved in the relevant decisions.

Moreover, Desfriches-Doria (2022) adds that what is at stake in these social and political dimensions of the management of the general and particular interest is, very often in contemporary controversies, the preservation of common goods: "The buildings, property, funds and other material resources involved in the exercise of this function are *res publica*, common goods" (Dewey 1927).

Controversies have been criticized for being asymmetrical or even being sometimes used as distractions by public authorities to redirect attention from more serious social problems (Chavot and Masseran 2016) and therefore creating obstacles to the democratic project and its deliberative function which should influence decision-making. Faced with plural conceptions of controversy and facing criticism placing limits on its study, it is important to circumscribe this concept with a theory whose advantages and pitfalls can be distinguished.

How can we approach controversy in the general theoretical framework of screen writings? First, it is quite obvious that mediation within this space is complex. To understand digital mediation, Yves Jeanneret and Emmanuel Souchier (2005) advanced the concept of screen writings that Jeanneret (2011) defines as a medium allowing documents of different nature and sources to be linked together in a single display.

Screen writings are therefore seen as particular types of computerized media which have an integrative function. The objective of integration in screen writings, for Jeanneret (2011), is to legitimize and reify cultural stereotypes. The signs used do not originate only from ancient cultural representations, but may also arise from current events.

But Christine Servais (2016, p. 10) argues that mediation is expressed today in "systems of a prescriptive, even coercive nature". Because this mediation is oriented by, if not hindered by issues of power, questions arise about conflict management, the plurality of norms and the conditions for sharing them, on the function of institutions and the role of policies in the management of society.

Michel Callon's actor–network theory (Callon and Ferrary 2006) allows us to extend the notion of actor to technical devices, to discourses, and at the same time to identify the meaning of the construction of affinity networks and to deduce from them the games of exclusion. But this principle of symmetry between actors initiated by David Bloor (1976) has been subject to strong criticism from certain constructivists, according to whom not all actors can have the same status in a relativist approach.

Chateauraynaud's sociological ballistics (2011a), with its descriptive and analytical approach to social conflicts, has the interest of analyzing the places where controversies emerge and their trajectories, of identifying the forces present and of questioning the development of themes. The interest of such an approach is that it authorizes a diachronic study of a single controversy, thus making it possible to investigate episodes made up of peaks, moments of calm and twists and turns or to grasp the development of reconfigurations of antagonistic groups and the evolution of their argumentation over time. For Chateauraynaud and Chavalarias (2017), the challenge of controversy is twofold: to *identify* a core of facts and to make explicit the presuppositions, theories or beliefs that determine the protagonists' interpretations and positions.

8.3. Methodological approaches

In the ICS field, as noted by Alexandre Coutant and Jean-Claude Domenget (2014), it has become customary to consider it productive to implement methodological pluralism; "this pluralism is exercised in a particular framework where the qualitative approach is widely dominant", note Françoise Bernard and Robert-Vincent Joule (2005, p. 186).

Our heuristic approach aims to answer the following questions: if methodological plurality is essential in ICS, what is the methodological combination best suited to exploring a controversy on YouTube and therefore proposing a suitable interpretation of it? Imbued with empiricist presuppositions, our method gives weight to the observation and occurrence of categories of controversy.

The corpus is established by varying the categories of actors as much as possible to avoid the pitfalls of media framing (Comby 2015), a concept which suggests that the echo found in the media of social debates and events is always oriented, involving a work of selection and prioritization of information for economic, political, strategic, etc. purposes. This reality of multiple issues also constrains the production of computerized media and leads the researcher to multiply sources of data by establishing a set of distinguishing criteria such as ideological, spatial, linguistic, temporal, etc. and to establish sub-corpora. By media, we mean here any device which serves as a medium for meaningful exchanges between people and which configures, to varying degrees, the form of these exchanges.

Left	Right	Liberal	Far right	Environmentalists
M	l'express	L'Opinion	minute	WWF
Libération	LE FIGARO	ouest france	VA	GREENPEACE
L'OBS	Le Point	Les Echos	C NEWS	Reporterre

Table 8.1. *Distribution of some actors in French media by ideological current. For a color version of this figure, see www.iste.co.uk/catellani/transitions.zip*

Several methods (screenshots, copy and paste, search engines, Hyphe) are used to collect data, each method having its advantages, but above all it is a question of guarding against volatility in the corpus, as was the case with YouTube hiding the "Dislike" counter on November 10, 2021.

To build the corpus, the starting point was controversies over the 2019 forest fires in the Amazon. As it is difficult to delimit a corpus on social media because of its shifting and labile nature, the solution proposed by Coutant and Domenget (2014) is to observe this corpus over a long period of time. In the case of the Amazon forest fires, the observation period extends over five years. In reality, this involved three years of synchronous observation from 2019 to 2021 and two years of asynchronous observation in order to understand the state of the controversy in 2017 and 2018. Ideological criteria, the technical nature of the videos, their visibility statistics and their length (short videos) are all relevant distinguishing elements. In total, this corpus includes 17 YouTube media videos and their comments, from which a sub-corpus of 108 websites was chosen. The Hyphe crawler was also used to extend this sub-corpus, allowing the discovery of new media actors in our controversy.

Links to the sub-corpus must be evoked either by the actors through hyperlinks produced on YouTube or through forms of direct or reported speech. We therefore start from these three scenarios to construct the sub-corpus:

– case of a link suggested by a commenter: "Man, just look at their president's political positions: https://fr.wikipedia.org/wiki/Jair_Bolsonaro";

– case of direct speech leading to the search for a source site: "Macron said: 'Our house is on fire'";

– case of reported speech leading to the search for a source site: "They talked about the tribes who live in the forests".

The texts extracted from the videos and especially the sub-corpus are analyzed in more depth using Prospero, an approach supported by Chateauraynaud and Chavalarias (2017, p. 153):

> These sub-corpora can then be the starting point for a more in-depth analysis of the content, in particular the narratives of events and forms of argumentation, using the Prospero suite. It is precisely a matter of contextualizing this analysis through the bifurcations identified in the groups concerned. Ultimately, this nesting of methods has the result of refining the masses of data so as to produce, on the one hand, broad categories providing a frame of interpretation and, on the other hand, indicators of certain regions of these masses of data, sufficiently circumscribed for a human interpreter to identify them and make a critical reading.

For the comments, given the complexity of the pattern of exchange, a manual analysis is required, with emphasis placed on the key comments for the most commented videos.

YouTube has a complex form of mediation which leads to an interrogation of complexity at the site level as well as at the video level. It is also a digital ecosystem including the YouTube application, the YouTube website and all the digital channels allowing YouTube to communicate with its targets, to improve the quality of its service and to guide its commercial strategies.

In this approach, the concept of screen writings, developed by Jeanneret and Souchier (2005), acts as a general theoretical framework in the construction of the composite methodological approach developed and defended here. It allows us to first interrogate the constituent elements of digital devices such as interfaces, modes and types of interactions, algorithms, types of visualizations, etc., in order to describe the medium of the corpus studied and its functioning and then to determine how this device produces, frames and constrains a certain mediation of the contents of the controversy and its configuration. Moreover, the tools used to analyze this controversy as software also help to highlight parts of these screen writings. Finally, they may also be the subject of a reflexive analysis of their contributions, possibilities and limits by considering their constituent elements as screen writings.

Jean Davallon and Yves Jeanneret (2004, p. 50) advocate screen writings as an alternative for going beyond graph tools to include "the technical dimension, the practice and the text" in the analysis. Investigating screen writings on YouTube in a context of controversy leads to a search for the different publishing actors in this controversy, to trying to unravel a fundamentally composite, labile and evolving computerized text, to investigating its architext and, more generally, the elements of YouTube's digital ecosystem that may influence the controversy. Screen writings are also extensions to the limits of Prospero, which cannot question the architexts, although these elements may influence the meaning and interpretation of the text (Jeanneret and Souchier 2005) and thus the configuration of the controversy.

In 2020, the petition site Avaaz criticized Google's algorithms for promoting the visibility of climate-skeptic websites[3]. In October 2021, Google decided to demonetize climate-skeptic sites[4]. These two cases underline the role of algorithms in the configuration of the controversy, an algorithm which increasingly tends to reduce the visibility of climate-skeptic voices. Analysis of screen writings also makes it possible to interrogate the editorial configuration of videos to discover that they are composed of elements from various media sources. Thus, from 2019 onwards, we increasingly see satellite images of forest fires being introduced into videos.

But an axiomatic approach[5] using crawler software (Hyphe) is also useful for macroanalysis of the morphology of groups. Here, the proof derives from a formal system and is deduced from an autonomous computational space. In order to refine the representation, the Gephi tool offers the possibility, with its numerous algorithms, of highlighting the desired categorizations. Indeed, exploration and representation always come together in cartography. This formalization provides an initial reading of the groups of media actors in the controversy based on their configurations and not their content (Mercklé 2004, p. 14). These crawls also allow us to determine which categories of actors cite which others. Who are the most and least cited actors? Which actors are isolated, and which form clusters of affinities? Which are the central actors?

---

3 Le Parisien, AFP (2020). Une ONG accuse YouTube d'orienter ses usagers vers des vidéos climatosceptiques. *Le Parisien*, January 16 [Online]. Available at: https://www.leparisien.fr/high-tech/une-ong-accuse-youtubed-orient-ses-users-vers-des-videos-climatosceptiques-16-01-2020-8237949.php.

4 Les Échos (2021). Réchauffement climatique: Google et YouTube sévissent contre la désinformation. *Les Échos*, 8 October [Online]. Available at: https://www.lesechos.fr/tech-medias/hightech/globalwarming-google-and-youtube-crackdown-against-desinformation-1353226.

5 An axiomatic system can usually be reduced to a formal theory. The mathematical theory of graphs aims to measure the structural properties of networks of relations using concepts, theorems, algorithms and modes of reasoning. It states that actors are characterized by their relationships rather than by their attributes. When this theory is used to model the Web, each web page is a vertex (or node) and each hyperlink is an arc between vertices.

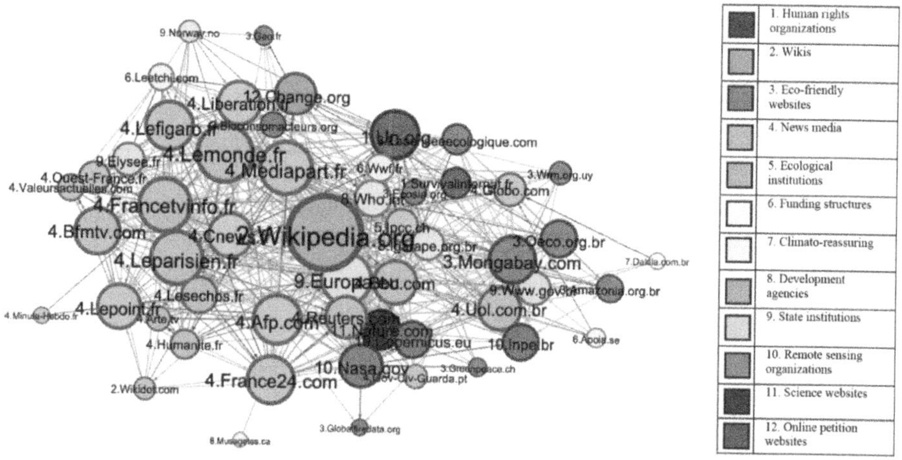

Figure 8.1. *Categorization by type of actor and by affinity based on links from 2019. For a color version of this figure, see www.iste.co.uk/catellani/transitions.zip*

On this graph, spatialized automatically by Gephi, certain actors are more or less isolated: this is the case of the far-right site Minute-hebdo and the climate-reassurist site Dakila. On the contrary, a relationship of homophily exists between certain sites such as Nature, Reuters, Copernicus and NASA, thus creating a cluster, while Wikipedia[6] presents itself as a central actor. Among this set of actors, some are places for concerted actions of commitment to the ecological cause: these are petition sites like Change.org. But with this description, the researcher cannot safely assume the reasons for the clusters, isolations and centralities, especially since the graph is only a snapshot of a network at a given moment.

Even though Jean-Christophe Plantin (2013, p. 68) rightly considers that "analyses of website graphs constitute an important tool for analysis of the creation of publics, mobilization of online actors and the deployment of controversies", these graphs above all highlight the key role of technical mediation in co-construction of interactions. A summary of the categories of actors involved in controversy in 2019 is given in Table 8.2.

6 The websites of the web giants are not included in crawls, except for Wikipedia due to its encyclopedia status.

Categories of actors			
Wikis	**Scientific sites**	**News media**	**Ecological sites**
State institutions	Development agencies	Human rights organizations	Climate reassurist sites
Ecological institutions	Actors involved in financing the protection of the Amazon forest	Remote sensing organizations	Online petition sites

Table 8.2. *Categories of actors at the media level in 2019*

This table of categories presents a variety of actors which already emphasizes the social character of this controversy. But, to use Bruno Latour's formula (2012), data visualizations are not in themselves sufficient, they are bases for moving on to deeper stages; otherwise, they would flatten the textuality and the weight of mediation through giving excessive space to the logistical dimension of communication. We can therefore agree that the map offered by the graph is in no way the same as the territory of the controversy.

This formalization of relationships between actors therefore ignores a large number of other online communication spaces such as petition artifacts or artifacts of showing support (sharing, liking, etc.). At the level of interpretation, "the instrument is suspected of conveying a mathematized conception of social processes moving away from a comprehensive approach to the interplay of actors and arguments, representations and practices which can only be grasped in situational context" (Chateauraynaud and Chavalarias 2017, pp. 137–138).

One of the most developed methods to overcome the limits of a structural analysis of networks of relationships is the socio-pragmatic approach, with the socio-informatics of controversies. Chateauraynaud (2006, p. 111) argues that with the Prospero suite, "to follow a process of investigation or a controversy is to see signs and modes of reasoning emerge, points of agreement and disagreement. Far from resembling a logical calculation of the traditional form, the processing of signs involves the continuous confrontation of connections and overlaps which bring together or divide authors-actors".

The objects sought here are linked to concepts of a "higher" logical level such as fictional beings, categories and collections, concepts which include lexico-semantic variations of the themes and make it possible to achieve an acceptable level of exhaustiveness and to examine, in a diachronic reading, the deployment over time of the cluster of themes and actors being studied. The other interest of this method is the possibility of adding on to argumentative or statistical operations. The researcher can query the text in the application itself and therefore relate the information obtained to the properties of the corpus, the forces present and series of successive events. Based on the videos denouncing the Amazon forest fires in 2019, by exploring the comments, we may identify the categories of actors in Table 8.3.

Forest fire disapprovers	Forest fire approvers
Vegetarians	Brazilian sovereignists
Vegans	Conspiracy theorists
Bio Consom'Acteurs	Climate skeptics
Environmentalists	Climate reassurers (thesis: global warming seen as cyclical)

Table 8.3. *Synoptic table of actors in the comments on the controversy in 2019*

We appreciate here the remark of Chateauraynaud and Chavalarias (2017, p. 153) who note that "social groups, whatever their definition, have vague and shifting boundaries". Indeed, someone may be a Brazilian sovereignist, a conspiracy theorist and a climate skeptic at the same time.

8.4. Conclusion

Construction of the methodology must be in agreement with the epistemological choices of the researcher in order to understand the way in which the latter discovers and constructs knowledge. With the use of measurement, collection, processing and analysis tools, the question of the positioning of the subjectivity of the researcher always returns to the center of the investigation. Beyond this epistemological and methodological approach to studying an ecological controversy on YouTube, the question of the investigability of this digital space always arises above all. A study of an online socio-technical controversy cannot be done without web analysis tools. But it is much more a matter of understanding how to use tools wisely to avoid bias and ensure plausible interpretations. We argue that a good qualitative approach can be made to the study of controversies on YouTube by combining exploration and representation using graph tools, a socio-informatics approach to controversies and screen writings.

8.5. References

Bernard, F. and Joule, R.-V. (2005). Le pluralisme méthodologique en sciences de l'information et de la communication à l'épreuve de la "communication engageante". *Questions de communication*, 7, 185–218. doi: 10.4000/questionsdecommunication.4647.

Bloor, D. (1976). *Knowledge and Social Imagery*. Routledge, London.

Callon, M. and Ferrary, M. (2006). Les réseaux sociaux à l'aune de la théorie de l'acteur-réseau. *Sociologies pratiques*, 13, 37–44. doi: 10.3917/sopr.013.0037.

Charaudeau, P. (2015). Qu'est-ce qu'une controverse ? CNRS, Laboratoire Communication et Politique, Paris [Online]. Available at: https://e-cours.univ-lr.fr/UNT/mediatisation/res/texte-charaudeau08b.pdf.

Chateauraynaud, F. (2006). Moteurs de (la) recherche et pragmatique de l'enquête. Les sciences sociales face au Web connexionniste. *Matériaux pour l'histoire de notre temps*, 82, 109–118. doi: 10.3917/mate.082.0109.

Chateauraynaud, F. (2007). La contrainte argumentative. Les formes de l'argumentation entre cadres délibératifs et puissances d'expression politiques. *Revue européenne des sciences sociales*, 45(136), 129–148. doi: 10.4000/ress.93.

Chateauraynaud, F. (2011a). *Argumenter dans un champ de forces. Essai de balistique sociologique*. Petra, Paris.

Chateauraynaud, F. (2011b). Sociologie argumentative et dynamique des controverses : l'exemple de l'argument climatique dans la relance de l'énergie nucléaire en Europe. *A contrario*, 16(2), 131–150. doi: 10.3917/aco.112.0131.

Chateauraynaud, F. and Chavalarias, D. (2017). L'analyse des grands réseaux évolutifs et la sociologie pragmatique des controverses : croiser les méthodes face aux transformations des mondes numériques. *Sociologie et sociétés*, 49(2), 137–161. doi: 10.7202/1054277ar.

Chavot, P. and Masseran, A. (2016). Controverse publique (sociologie des sciences). *Publictionnaire. Dictionnaire encyclopédique et critique des publics* [Online]. Available at: http://publictionnaire.huma-num.fr/notice/controverse-publique-sociologie-des-sciences [Accessed 7 May 2018].

Comby, J. (2015). La politisation en trompe-l'œil du cadrage médiatique des enjeux climatiques après 2007. *Le Temps des médias*, 25, 214–228. doi: 10.3917/tdm.025.0214.

Coutant, A. and Domenget, J.-C. (2014). Understanding users' expectations for recommender systems: The case of social media. In *Recommender Systems*, Kembellec, G., Chartron, G., Saleh, I. (eds). ISTE Ltd, London, and John Wiley and Sons, New York.

Dagicour, O. (2020). Géopolitique de l'Amazonie. *Politique étrangère*, 135–146. doi: 10.3917/pe.201.0135.

Davallon, J. and Jeanneret, Y. (2004). La fausse évidence du lien hypertexte. *Communication & langages*, 140, 43–54. doi: 10.3406/colan.2004.3266.

Desfriches Doria, O. (2022). La modélisation cartographique des controverses, un outil au service de l'esprit critique ? *Approches Théoriques en Information-Communication (ATIC)*, 5, 47–77. doi: 10.3917/atic.005.0047.

Desfriches Doria, O., Dessinges, C., Ihadjadène, M. (2018). Controverses et maîtrise de l'information : analyse et propositions méthodologiques. *Technologie et innovation*, 18(3), 18 [Online]. Available at: https://www.openscience.fr/IMG/pdf/7_doria_ok.pdf.

Dewey, J. (1927). *Le public et ses problèmes*, Zask, J. (trans.). Gallimard, Paris.

Elster, J. (2005). L'usage stratégique de l'argumentation. *Négociations*, 4(2), 59–82. doi: 10.3917/neg.004.0059.

Jeanneret, Y. (2011). L'écrit d'écran : lire, écrire et un peu davantage. In *Y-a-t-il (vraiment) des technologies de l'information ?*, Jeanneret, Y. (ed.). Presses Universitaires du Septentrion, Villeneuve-d'Ascq, 147–189 [Online]. Available at: http://books.openedition.org/septentrion/13906.

Jeanneret, Y. and Souchier, E. (2005). L'énonciation éditoriale dans les écrits d'écran. *Communication & langages*, 145, 3–15. doi: 10.3406/colan.2005.3351.

Latour, B. (2012). Biographie d'une enquête. À propos d'un livre sur les modes d'existence. *Archives de philosophie*, 75, 549–566. doi: 10.3917/aphi.754.0549.

Mercklé, P. (2004). *Sociologie des réseaux sociaux*. La Découverte, Paris.

Plantin, J.-C. (2013). Qu'y a-t-il à côté d'un graphe de sites web ? Analyse de réseaux sociaux et pratiques de synchronisation en ligne. *Communication & organisation*, 43, 59–70. doi: 10.4000/communicationorganisation.4127.

Rennes, J. (2016). Les controverses politiques et leurs frontières. *Études de communication*, 47, 21–48. doi: 10.4000/edc.6614.

Scotto d'Apollonia, L. (2014). Les controverses climatiques : une analyse socioépistémique. PhD Thesis, Université Paul Valéry-Montpellier III, Montpellier [Online]. Available at: http://www.theses.fr/2014MON30024 2014/10/14.

Servais, C. (2016). *La médiation. Théorie et terrains*. De Boeck Supérieur, Brussels. doi: 10.3917/dbu.serva.2016.01.

Servigne, P. and Stevens, R. (2015). *Comment tout peut s'effondrer. Petit manuel de collapsologie à l'usage des générations présentes*. Le Seuil, Paris.

Souchier, E., Candel, É., Gomez-Mejia, G., Jeanne-Perrier, V. (2019). *Le numérique comme écriture. Théories et méthodes d'analyse*. Armand Colin, Paris.

Venturini, T. (2012). Building on faults: How to represent controversies with digital methods. *Public Understanding of Science*, 21(7), 796–812. doi: 10.1177/0963662510387558.

9

Sustainability Transitions in Tension: Sustainability Transitions in the Discourse of Eco-Shaming

9.1. Sustainability transition: what's in a name?

Nowadays, calls for a sustainability transition are common among researchers, citizens, policymakers and activists. The concept of a sustainability transition refers to "a radical transformation towards a sustainable society" (Grin et al. 2010, p. 1), but does not specify the content or form of this transformation, nor the way in which it takes place or is expected to take place (Audet 2014, 2016; Silva and Stocker 2018; Feola and Jaworska 2019). Consequently, the notion of a sustainability transition refers to multiple and divergent visions of societal transformation (Audet 2014; Luederitz et al. 2016). Understanding the different interpretations of the concept is important when, faced with increasing social and environmental degradation, we aspire to advance and guide this transformation in a precise and never neutral direction. Several specific interpretations of sustainability transitions have already been analyzed, such as for "transition towns" (Barry and Quilley 2009; Dryzek 2021). The objective of this chapter is to account for the particular interpretation of a sustainability transition in the discourse of eco-shaming.

The discourse of eco-shaming has spread with the media coverage of Greta Thunberg and her appeals of "how dare you?" This discourse is characterized by practices of eco-shaming that (re)produce it, i.e. practices of shaming someone for behavior which has a negative impact on the environment; this discourse presents a normative and moral approach to environmental issues. The notion of discourse refers to "a shared way of apprehending the world […] [which] conditions the way

Chapter written by Kimberley VANDENHOLE.

in which we define, interpret and address environmental issues" (Dryzek 2021, pp. 9–12) and thereby encompasses a particular vision of sustainability transitions. Thus, dissemination of the eco-shaming discourse not only challenges our ways of perceiving the environment, but also modifies our understanding of sustainability transitions.

The specific interpretation of sustainability transitions in the discourse of eco-shaming presented below results from a discourse analysis whose conceptual and methodological framework is explained below. The chapter ends with a positioning of the discourse of eco-shaming within the landscape of interpretations of sustainability transitions.

9.2. Analysis of the eco-shaming discourse and its notion of sustainability transitions

Giving meaning to sustainability transitions through discourse favors the mobilization of discourse analysis as a theory and a methodology (see Dryzek 2021, pp. 9–13). Starting from the existence of multiple environmental discourses encompassing distinct conceptions of sustainability transitions, to understand these conceptions requires discourse analysis.

The analysis conducted in the context of the eco-shaming discourse consisted in a study of 2,155 textual documents through a rigorous and extensive coding procedure that involved repeated and systematic examination of the data using the NVivo12 program. As the analysis is limited to Belgium, the documents include newspaper articles from *Het Nieuwsblad*, *De Standaard*, *De Morgen*, *Le Soir*, *L'Avenir* and *La Libre Belgique*; written parliamentary questions from the federal, Brussels, Flemish and Walloon Parliaments; environmental movement newsletters from Natuurpunt, WWF, Greenpeace and Friends of the Earth. Analysis of documents from three different fields allowed coding of the discourse of eco-shaming as it is constructed, expressed and disseminated by and for different actors and in different contexts. The chosen documents contain one or more of the words "climate", "environment", "climate change", "global warming", "shame" (in Dutch or French) in their title and were collected from databases online. They were all issued in Belgium between January 2018 and May 2021.

Coding consisted of inductive detection of utterances, meaning textual pieces that indicate, contain or state a storyline or a practice of the eco-shaming discourse. As a shorthand for discourse – these are abbreviated narratives that conceal and condense the complex discursive structure of discourse – storylines clearly express the problematization and particular resolution of environmental problems supported by the discourse (Hajer 2006). Storylines comprise three frames: diagnostic frames describe the past and contemporary state of the system and identify problems associated with it; prognostic frames describe a predicted or desired future; motivational frames describe how to move from the current state to the desired state (Snow and Benford 1988). A sustainability transition may be characterized by three phases: a given state of the system, a process of change and a new state of the system (de Haan and Rotmans 2011). The particular articulation of this transition process is called a "transition pathway". The phases of transition correspond to the discourse frames mentioned above. Therefore, the storylines of the eco-shaming discourse inform us about the transition pathway imagined and promoted by the discourse, while the practices of the discourse make visible this specific interpretation of a sustainability transition by translating it into social acts.

9.3. Sustainability transition in the eco-shaming discourse

Sustainability transition is "a response to a number of persistent problems facing contemporary modern societies" (Grin et al. 2010, p. 1). The way that it is conceived thus depends on the problems identified, i.e. on the definition used of what constitutes a problem. In the eco-shaming discourse, the storyline summarized by the phrase "there is commonly an imbalance between the natural world and the human world" defines the problems to be solved by a sustainability transition. This storyline is based on three ideas: firstly, "the natural world is distinct from the world of humans"; at the same time, secondly, "the two worlds exert a mutual influence" and third, "the natural world enables the human world through ecosystem services and natural capital". The diagnostic frame for imbalance is anthropocentric, as illustrated by the following expression: "We all find it shameful to see this industry developing into a French fry factory that is harmful to agriculture, local residents and the health of consumers!" (Les amis de la terre 2020). This frame refers to the notion of "excess": the use of the natural world by the human world is excessive from the moment it creates a disadvantage for the human world, as argued by Greenpeace (2018): "We all eat too much meat [...] the current level of meat production [...] has enormous consequences for humans, animals and the environment". The diagnostic frame evoked by the storyline of imbalance limits the conception of a sustainability transition in the eco-shaming discourse to an anthropocentric "transition" (pertaining to the human world) and thus can be seen as weakly radical.

The storyline of imbalance is complemented by the storyline that "some people do not take responsibility for the natural world". According to the eco-shaming, discourse, persistent problems result from part of the population not taking responsibility. This diagnostic frame is based on the following ideas: "the natural world is a public good" and "each person has responsibilities towards the public good", as stated in the Brussels Parliament (2020): "[...] the responsibility of everyone in the conservation of a common resource [water]" and leads to the prognosis summarized by the storyline of "the tragedy of the commons". The diagnostic frame of responsibility refers to agency (being responsible presupposes a specific capacity to act). A sustainability transition, according to the eco-shaming discourse, is therefore primarily located at the level of people's agency.

The notion of responsibility mobilized by the eco-shaming discourse is summarized by the storyline of "common but differentiated responsibilities", in the sense that responsibility for the public good belongs to humanity as a global entity, but the content and scope of this responsibility differ according to people's social role, as formulated in the Flemish Parliament (2020): "[...] but responsibility is shared by everyone. Users may be required to change their behavior and their habits, producers their economic models". Each social identity has its own role to play in maintaining the public good. This specific notion of responsibility implies recognition of agency as being shared by all, but different for all, and thus projects a sustainability transition supported by everyone consisting of multiple, particular, and non-uniform actions. For example, given the contribution of air traffic to several environmental problems, sustainability transition in the eco-shaming discourse might refer equally to the responsibility of political decision-makers to establish a kerosene tax as to the responsibility of travelers to reduce their air travel and to the responsibility of the aeronautical industries to increase the energy efficiency of planes. The non-specific nature of the notion of responsibility advanced by the eco-shaming discourse creates a very broad prognostic framework for completion of the transition.

The notion of responsibility, at the heart of the eco-shaming discourse, has an inherent moral and normative character: it transforms agency as the capacity to act into a duty to act and in fact a personal and societal expectation. The eco-shaming discourse uses a motivational frame of morality and normativity as synthesized in the storylines "every person must take responsibility for the natural world and that of humans", "contemporary society insufficiently recognizes the impacts that certain people have on others" and "people who do not take responsibility are bad people", as formulated in *De Morgen* (2018): "Building a single-family home now is criminal". According to the eco-shaming discourse, a sustainability transition passes through two channels. The first is strengthening of the social and moral standards in force, that is requiring that responsibility be taken, for example, by strengthening carbon offsets during plane travel or in the case of the Walloon Parliament (2021): "The Minister of the Environment has a responsibility in these issues of illegal transfers of waste" or in *Het Nieuwsblad* (2018): "It would be a good signal for packaging producers to finally take responsibility". The second channel is the implementation of changes to social and moral norms: this means creating responsibilities, for example, by making air travel socially and morally undesirable or in the case of the Walloon Parliament (2021): "The region is therefore responsible for the growth of this company to the obvious detriment of the environment and working conditions [...]".

The approach of the eco-shaming discourse towards reinforcement, imposition and construction of responsibility (the desired prognostic frame) is also normative and moral, promoted by the storyline "it is appropriate to resort to eco-shaming practices to impose a sense of responsibility". It aims to create a sense of both intrinsic (moral) and extrinsic (social, normative) responsibility, as in the Walloon Parliament (2018): "Is it not lacking a sense of political responsibility not to set up within Wallonia's public service services concentrating on the health-environment link?" rather than an exclusively extrinsic sense of responsibility (e.g. motivated by legislation). Eco-shaming practices are conceived as a lever for transitions. They may be used by anyone, thereby constructing humanity as an actor in the transition. The moral and normative approach of eco-shaming discourse towards a sustainability transition is well reflected in eco-shaming practices, which focus more on the lack of responsibility than on the intensity of the environmental impact of behaviors. The motivational frame posed by eco-shaming practices is mainly defined in negative terms, as shown in the storylines that speak of "crisis" and "resistance". The call is for immediate actions of resistance (against a predicted future) rather than emancipation (towards a desired future).

Diagnosis: past and current status of the problematic system	**Motivation: models of change**	**Motivation: state of the intermediate system**	**Motivation: models of subsequent change**	**Prognosis: desired and imagined state of the future system**
An anthropocentric imbalance between the natural world and the world of humans	Resistance	General recognition of the environment as a public good	Resistance	An anthropocentric balance between the natural world and the world of humans

<table>
<tr><td></td><td>Re-establishing balance between the natural world and the human world</td><td></td><td>Re-establishing balance between the natural world and the human world</td><td>Every human being assumes responsibility for protecting the commons (i.e. the environment)</td></tr>
<tr><td rowspan="2">Some people do not take responsibility for protecting the commons (i.e. the environment)</td><td>Eco-shaming practices</td><td rowspan="2">Changing social norms and normative, moral principles</td><td>Eco-shaming practices</td><td>Eco-shaming practices</td></tr>
<tr><td>Every human being has their role to play</td><td>Every human being has their role to play</td><td>Shared but differentiated responsibilities</td></tr>
<tr><td rowspan="2">Undifferentiated responsibilities entrusted to a small number of actors (notably citizens)</td><td>Strengthening the social and moral norms in force (enforcing responsibility)</td><td rowspan="2">General recognition of shared but differentiated responsibilities in relation to the commons</td><td>Strengthening the social and moral norms in force (enforcing responsibility)</td><td>New social norms</td></tr>
<tr><td>Changes in social and moral norms (generating responsibilities)</td><td>Changes in social and moral norms (generating responsibilities)</td><td>Following normative, moral and social principles</td></tr>
</table>

Table 9.1. *A sustainability transition according to the eco-shaming discourse: transition pathways and storylines. Based on de Haan and Rotmans (2011)*

9.4. One sustainability transition among others

So far, the text has made explicit the particular notion of a sustainability transition in the eco-shaming discourse (Table 9.1). In the following, this notion is compared to the sustainability transitions presented by other environmental discourses (see Dryzek 2021), in order to further emphasize its specificity.

At the heart of different interpretations of sustainability transitions are distinct notions of the levers and actors of transitions. In the eco-shaming discourse, the levers of transitions are moral, social and normative principles and practice, whereas, for example, they are market measures in the discourse of economic rationalism, political changes in the discourse of green politics and technological innovations and transition management practices in the discourse of ecological modernization and the Promethean discourse (see Dryzek 2021). As levers of the transition, the principles of responsibility and practices of eco-shaming allow anyone to contribute to a sustainability transition. Thus, the eco-shaming discourse does not limit actors of the transition to specific groups as do, for example, the discourses of green politics (political leaders), administrative rationalism (experts) and green consciousness (enlightened postmaterialists). Even though everyone is called to contribute to a sustainability transition in the eco-shaming discourse, not all are called in the same way (as expressed by the notion of shared but differentiated responsibilities), as opposed, for example, to the discourses of democratic pragmatism and green consciousness. "Transition as understood by the eco-shaming discourse is incumbent on humanity as a whole and is not imposed by political or economic leaders (as in the discourses of green politics and economic rationalism) or by experts (as in the discourse of administrative rationalism)".

Designating different levers and actors of the transition compared to other discourses, nevertheless, the discourse of eco-shaming fundamentally supports a vision of a sustainability transition similar to that shared by most environmental discourses: a sustainability transition serves the human world, it is ultimately anthropocentric. As a result, it is less radical than the sustainability transition advanced by discourses of the deep green type.

9.5. Conclusion

Nowadays, calls for a sustainability transition are common among researchers, citizens, policymakers and activists. The undefined nature of the concept allows several actors to cooperate and unite around a common objective. However, different interpretations of a sustainability transition may jeopardize the transition process. This is where knowledge of these interpretations seems particularly interesting. This chapter contributes to the expansion of such knowledge through its account of the specific interpretation of a sustainability transition advanced by the emerging discourse of eco-shaming.

9.6. References

Audet, R. (2014). The double hermeneutic of sustainability transitions. *Environmental Innovation and Societal Transitions*, 11, 46–49. doi: 10.1016/j.eist.2014.02.001.

Audet, R. (2016). Transition as discourse. *International Journal of Sustainable Development*, 19(4), 365–382. doi: 10.1504/IJSD.2016.10001427.

Barry, J. and Quilley, S. (2009). The transition to sustainability: Transition towns and sustainable communities. In *The Transition to Sustainable Living and Practice*, Leonard, L. and Barry, J. (eds). Emerald Group Publishing, Bingley, 1–28.

Dryzek, J.S. (2021). *The Politics of the Earth*. Oxford University Press, Oxford.

Feola, G. and Jaworska, S. (2019). One transition, many transitions? A corpus-based study of societal sustainability transition discourses in four civil society's proposals. *Sustainability Science*, 14, 1643–1656. doi: 10.1007/s11625-018-0631-9.

Grin, J., Rotmans, J., Schot, J. (2010). *Transitions to Sustainable Development. New Directions in the Study of Long-Term Transformative Change*. Routledge, New York.

de Haan, J. and Rotmans, J. (2011). Patterns in transitions: Understanding complex chains of change. *Technological Forecasting and Social Change*, 78(1), 90–102. doi: 10.1016/j.techfore.2010.10.008.

Hajer, M. (2006). Doing discourse analysis: Coalitions, practices, meaning. In *Words Matter in Policy and Planning*, van den Brink, M. and Metze, T. (eds). Koninklijk Nederlands Aardrijkskundig Genootschap, Utrecht, 65–74.

Luederitz, C., Abson, D.J., Audet, R., Lang, D.J. (2016). Many pathways toward sustainability: Not conflict but co-learning between transition narratives. *Sustainability Science*, 12, 393–407. doi: 10.1007/s11625-016-0414-0.

Silva, A. and Stocker, L. (2018). What is a transition? Exploring visual and textual definitions among sustainability transition networks. *Global Environmental Change*, 50, 60–74. doi: 10.1016/j.gloenvcha.2018.02.003.

Snow, D. and Benford, R. (1988). Ideology, frame resonance, and participant mobilization. *International Social Movement Research*, 1, 197–219.

10

Discursive Controversies Around Electric Vehicles in France: Between Ecological Transition and Doubt

10.1. Introduction

Among the subjects linked to the measures required to deal with climate change, that of transport is one of the most iconic; on a daily basis, everyone uses transport that is more or less active, public or low in CO_2 emissions. At the heart of proposed solutions to combat climate change, electric vehicles are well-established: we know their effects on urban air quality (Brady and O'Mahony 2011; Nanaki and Koroneos 2016) and certain automobile manufacturers have long been invested in the subject, such as Renault (Castellano et al. 2011), with the support of public authorities (Calef and Goble 2007).

The issue is at the same time industrial and linked to public policies; originally conceived to fight urban air pollution, the electric car is now cited as a potential solution to fight climate change, along with public transport and active (soft) transport. However, it is far from unanimous; several studies have demonstrated the pollution linked to the sector, from the production of batteries to their recycling, not to mention the price (Nykvist and Nilsson 2015). It is precisely to these controversies that this proposal is devoted, especially in a hypernarrative context that is particularly prolific on social networks (Wagener 2020).

Chapter written by Albin WAGENER.

10.2. Theoretical and methodological framework

To study this discursive complex, we avail ourselves of discourse analysis. To do this, we consider the specificities of environmental discourses (Alexander 2009; Biros 2014), particularly in their activist dimensions (Carlino and Stein 2019), as well as the role of the construction and use of corpora in discourse analysis (Baker 2006). In other words, our work is rooted in a theoretical framework based on the following elements:

– Discourse is a social object, which uses lexical and semantic possibilities to give color to shared and shareable narratives, which influence the social world as much as the social world influences discourse.

– The study of discourse provides access to the way in which social representations are produced, put into circulation, and received.

– Creation of a corpus makes it possible to obtain a statistically significant set of data, in order to measure the frequency of certain occurrences and lexical collocations.

This theoretical framework, largely focused on tooled discourse analysis, especially as applied to corpora, is particularly relevant here in the context of electric vehicles. Indeed, we hypothesize that the representations circulating about these vehicles generally oscillate between two poles: on the one hand, the fact that the electrical aspect allows a response to the question of pollution blamed on petrol cars; on the other hand, the fact that the life cycle of electric batteries, in particular, constitutes an ecological mismanagement ultimately just as ecocidal as that of petrol vehicles.

Consequently, the aim of this study is to examine how representations about electric vehicles circulate, by means of a tooled discourse analysis applied to the corpus, and based on texts published in France which are sufficiently varied and open to guarantee a representative heterogeneity of lexical data (Garric 2012). This heterogeneity must in turn make it possible to embrace a sufficiently broad set of discourses, and therefore of social representations, about electric vehicles. In addition, these corpora are constructed with the argumentative deviations of climate-skeptic discourses in mind (Wagener et al. 2021) which also feature in some of the discourse presented. Thus, our corpus consists not only of media articles on the subject of electric vehicles, but also a collection of communicational occurrences collected on several social networks, namely Twitter, Facebook, YouTube and Reddit. The textual corpus is composed as follows, thanks to a collection of data over one year, from May 1, 2020 to May 1, 2021:

– comments from the 10 Facebook pages showing the highest engagement rate (3,500 comments);

– comments from the 10 YouTube pages showing the highest engagement rate (2,000 comments);

– comments from the 10 Reddit pages showing the highest engagement rate (360 comments);

– the 100 Tweets showing the highest engagement rate;

– the 10 most shared media articles on social networks (including Reporterre (for two articles), then France 3, the Huffington Post, RFI, Auto Propre, Le Parisien, LCI, Clubic and FrAndroid).

Posts were selected using the CrowdTangle tool, which finds the posts with the highest engagement rate (comments and shares) on social networks; the selected pages are thus varied, both in terms of content and format, but have the common feature of featuring electric vehicles as a topic of discussion.

Our approach is at the crossroads of discourse analysis and digital humanities, insofar as we used several tools to extract our corpus:

– Mediacloud, which allows extraction of the most shared media articles on social networks;

– CrowdTangle, which allows extraction of the most successful Facebook posts;

– Facepager, which has the advantage of allowing extraction of data from several social networks in an automated manner.

For analysis of this corpus, we used other tools to enable a study of the data:

– IRaMuTeQ, which offers analysis by semantic cluster and by relationships of meaning between different elements of the corpus;

– Gephi, which allows formatting of different visualizations.

These tools are not neutral for our analysis: they allow us to grasp the regularities and specificities of the lexical and semantic data contained in the corpus, to better grasp the communicational and argumentative axes.

Obviously, a tooled analysis is enlightening, but it is not enough to identify the specificities of the corpus: we therefore complement this initial analysis with what we call the return to the corpus, which allows us to better grasp argumentative expressions, particularly in their dimensions of mediation of representations (Kopytowska 2015) and pragmatic and cognitive weight (Maillat 2013).

In fact, the analysis we propose is systemic (Wagener 2019): it aims to identify the sets of arguments hidden at the heart of the controversy, and particularly to understand what motivates the arguments of those who adopt electric vehicles, or those who reject this solution.

More broadly, this study aims to re-establish the example of an ecological and socio-technical controversy not only at the heart of social network activity, but also in connection with the necessary cultural and domestic upheavals caused by these changes in habits (Poupon et al. 2017).

In addition, this study must allow us to understand the interests of groups and individuals who defend certain arguments rather than others, but also to check for any argumentative diversity depending on the media and the different social networks used by citizens and political or economic actors.

10.3. Corpus

The overall treatment of these elements makes it possible to obtain, as a first step, a word cloud illustrating the most frequent occurrences within the corpus, consisting of comments responding to the articles. This word cloud constitutes a visualization that provides us with the first keys of entry into the corpus, insofar as the most frequent terms may be used as representational and discursive indices. Obviously, such a visualization is indicative and is only a preliminary step in analysis of the corpus.

To identify the argumentative and discursive logics at work in the 10 main articles, these were extracted and encoded for processing in IRaMuTeQ. We thus have access to the representations in circulation concerning electric vehicles; they constitute real signposts towards the narrative logics to be identified. This mapping of the lexemes circulating in the public space about electric vehicles is all the more relevant as Internet users travel from platform to platform and from information source to information source. Here we find, in particular, an emphasis on the battery, which gives us an indication of what is to follow below.

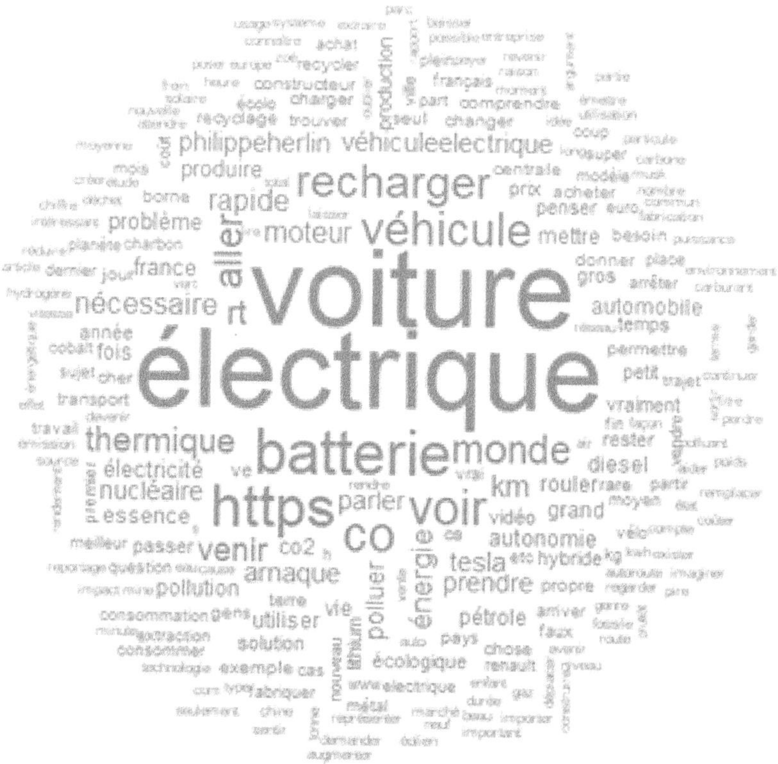

Figure 10.1. *Word cloud (source: IRaMuTeQ). For a color version of this figure, see www.iste.co.uk/catellani/transitions.zip*

The analysis of similarities, mirroring the word cloud (Figure 10.1), shows how the most commonly used terms in the corpus relate to each other and are accompanied by a set of words which already show the semantic and lexical links between certain ideas and arguments present within the corpus. Through this analysis, which allows us to offer a visualization of the networking of lexical collocations within the corpus, we can go beyond a simple word cloud. This is particularly enlightening, because we can see a separation between cars (as a means of transport) with their CO_2 emissions and their technical characteristics and the electrical dimension (with recharging, dependence on nuclear energy and the humanitarian dimension of the use of children to extract minerals). Thus, the battery finds itself isolated from the other two sets, because of the treatment of its life cycle, from production to recycling.

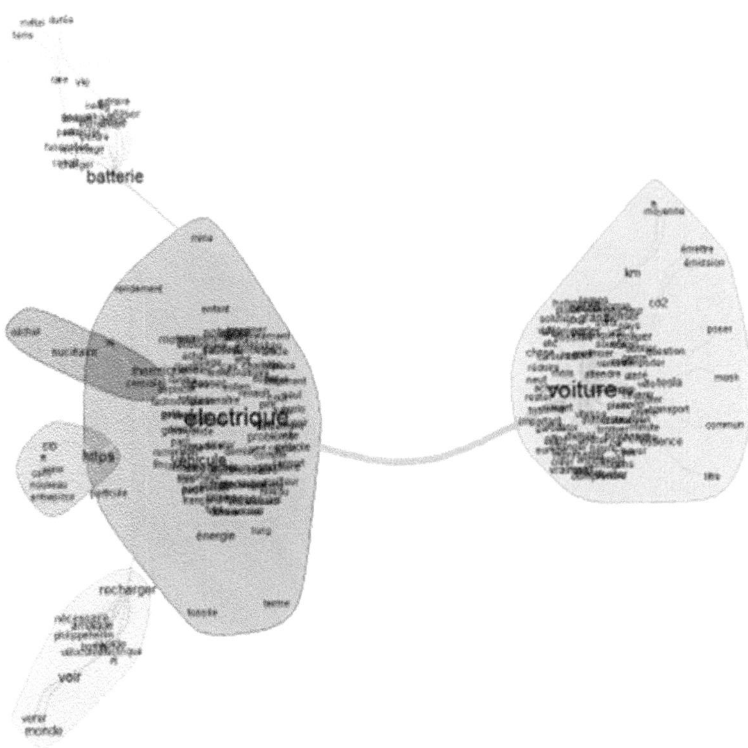

Figure 10.2. *Similarity analysis (source: IRaMuTeQ). For a color version of this figure, see www.iste.co.uk/catellani/transitions.zip*

To better understand the links between terms within the corpus and to deepen the analysis enabled by the analysis of lexical and semantic universes (Figure 10.2), it is important to understand how semantic elements are articulated into distinct groups, which operate according to a precise semantic and argumentative logic. With this in mind, one of the most interesting features of IRaMuTeQ is its ability to isolate the major themes by class (Figure 10.3).

It is interesting to note that from an argumentative point of view, across the entire corpus of comments, we observe four classes, two of which are particularly significant:

– class 1 (in red, 40%) dealing with the technical performance and purchasing criteria of electric vehicles;

– class 2 (in green, 36%) dealing with the ecological impact of electric vehicles;

– class 3 (blue, almost 10%) showing the engagement and judgment of the values of Internet users, focusing mainly on the ecological impact: this is not trivial and shows that the ecological question provokes a specific engagement;

– class 4 dealing with the digital environment of Tweets.

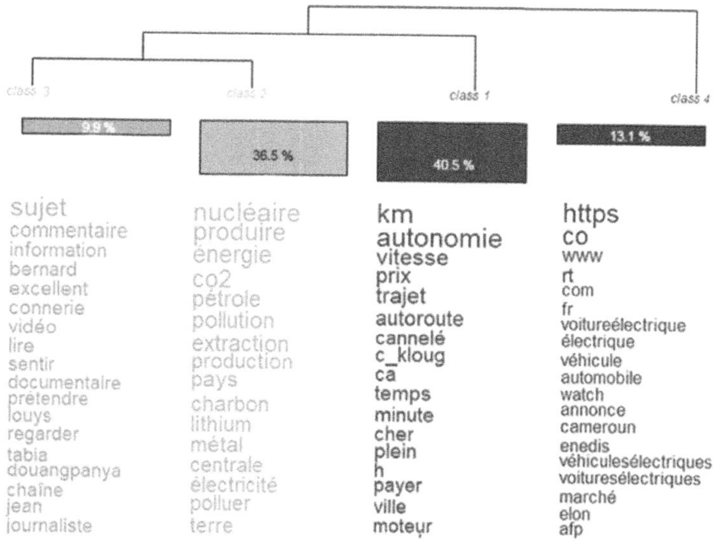

Figure 10.3. *Dendrogram analysis of the themes of the corpus (source: IRaMuTeQ). For a color version of this figure, see www.iste.co.uk/catellani/transitions.zip*

To go further, it is also important to understand the links and positionings between the thematic classes isolated by IRaMuTeQ (Figure 10.3), in order to understand the logic of proximity or distance between the different themes present in the corpus – which makes it possible to identify, among other things, the axiological values of the discursive elements being linked together. This functionality of IRaMuTeQ allows visualization of the mutual positioning of thematic classes, which offers a useful map to understanding the argumentative logic within the corpus.

Thus, putting the links between the classes into perspective (Figure 10.4) shows two things: first, a clear distance between classes 4 (the digital universe of corpus production) and 1 (the performance of electric vehicles).

Then, and clearly, an almost total interlinking of classes 2 and 3, showing that the involvement and attention of Internet users are particularly acute when the theme of electric vehicles is approached from the point of view of environmental impact and battery pollution, nuclear energy consumption or the conditions of mineral extraction.

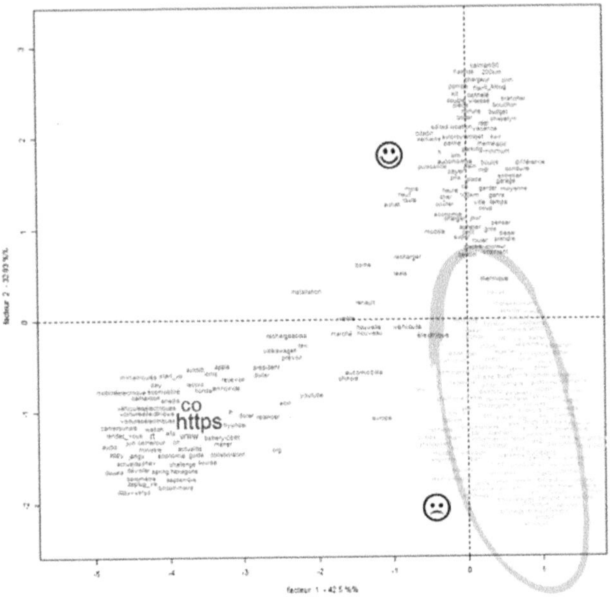

Figure 10.4. *Factorial analysis of correspondences (source: IRaMuTeQ). For a color version of this figure, see www.iste.co.uk/catellani/transitions.zip*

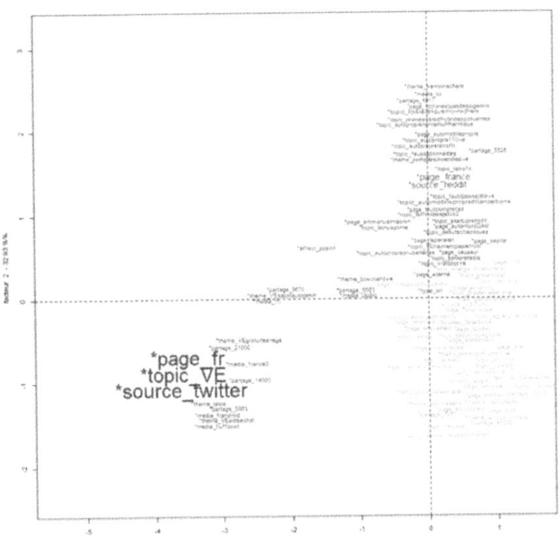

Figure 10.5. *Factorial analysis of the correspondences of corpus variables (source: IRaMuTeQ). For a color version of this figure, see www.iste.co.uk/catellani/transitions.zip*

To better understand how these relationships are constructed within the corpus and to understand the specificities of the texts used, metadata (or the sources and characteristics of the texts) may also be studied, to refine the lessons of our inquiry (Figure 10.5).

Analysis of the metadata of the corpus, or the variables distinguishing the type of social network, the subject or the affect, shows Twitter with a very different positioning with relation to all the other sources of the corpus; this is due to the high volatility of the network, its functioning and retweets, which deserve a separate treatment. We also perceive a positioning of Reddit within class 1, i.e. around the technical performance of electric vehicles.

On the contrary, the other sources show a more critical positioning, especially concerning the conditions of production and recycling of electric batteries (both from an ecological and humanitarian point of view).

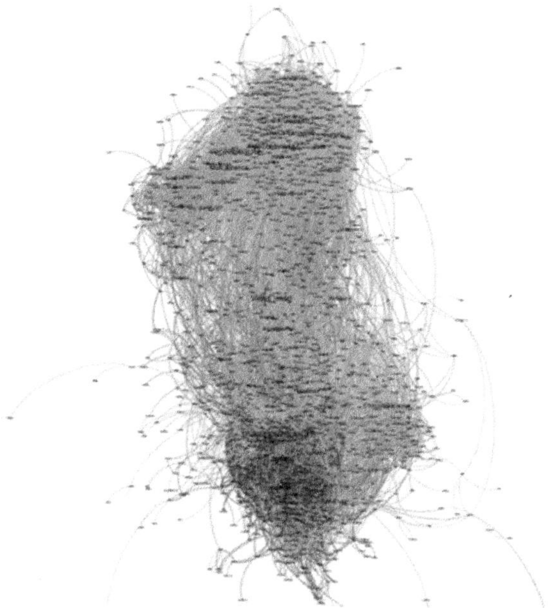

Figure 10.6. *Representation of conversational ecosystems on Twitter (source: Gephi via Pulsar). For a color version of this figure, see www.iste.co.uk/catellani/transitions.zip*

To complement our corpus analysis, and to better understand the enunciative positions which fuel the representations in circulation about electric cars, we also wanted to use Twitter – not by including tweets in the corpus strictly speaking, but to try to understand the logic of enunciative postures which provide relatively positive or negative representations of electric vehicles (Figure 10.6). Such a complementary study sheds further light on the results interpreted using IRaMuTeQ. Concerning electric vehicles, the network on Twitter is very dense, although there is a distinction between thematic groups regarding treatment of the subject.

The argumentative logics on display have their own particularities, depending on whether we are towards the north of the diagram (in purple) or towards the south-east (in orange); these differences are due not only to the socio-political galaxies of the accounts, but also to the narratives that fuel support for or opposition to the solution offered by electric vehicles. The following can thus be distinguished:

– in the northeast, in purple: the sovereignist and conspiracy galaxy, unfavorable to electric vehicles because of the theme of deindustrialization;

– to the northwest, in green: the environmentalist galaxy, relatively unfavorable to electric vehicles, due to a preference for public or active transport;

– to the west, in turquoise: a skeptical ecological galaxy, particularly linked to the media treatment of the subject;

– to the south, in black: a dense network of support for electric vehicles, linked in particular to accounts supporting technological innovations;

– in the center, in gray: a network of media relatively positive on the issue;

– in the east, in orange: an official political discourse from the government, positively engaged on the subject.

These visualizations show one thing: there are extremely diverse narratives on the subject of the electric car. It is this narrative analysis that is important for understanding the thematic and argumentative contents of the corpus.

10.4. Narrative analysis

In the light of these analyses, and through a return to the corpus which allows us to use linguistic materialities as a basis for understanding the stories in circulation, it is possible to distinguish several specificities.

To begin with, there are positive stories positioned in favor of electric vehicles, in line with majority technophile or media discourse. Here, we find the following arguments:

– "The electric car helps reduce urban pollution": even though opponents of electric vehicles explain that the best way to reduce urban pollution is to use public transport or active modes, the fact remains that the use of these vehicles allows a reduction in the proportion of exhaust gases in cities, and therefore the production of CO_2.

– "The electric car is less polluting than internal combustion vehicles": this positioning emphasizes the reduction of urban pollution, but makes it possible to mobilize technical arguments in terms of energy performance or quite simply technical innovation. This narrative links engineering and innovation in industrial technology.

– "Retrofitting allows transformation of an internal combustion vehicle into an electric vehicle": this narrative plays on the theme of an authentic return to mechanical craftsmanship, in a do-it-yourself spirit, which often accompanies ecological arguments. In this sense, it involves recycling a polluting object into a virtuous object.

There are also stories positioned relatively against the use and circulation of electric vehicles, along the following argumentative lines:

– "Electric batteries are polluting and represent a humanitarian cost": this argument is one of the most widespread, particularly following a documentary broadcast on Arte entitled "The hidden face of green energies", which was based on reports and survey work documenting both the conditions of mineral extraction, and also the entire life cycle of batteries, from manufacturing to recycling issues.

– "The electric car is not a sustainable solution to the transport issue": this argument compares the electric vehicle to other modes of transport, such as public transport (car, bus, train) or active modes (bicycle, scooter). On the contrary, this argument avoids the situation of peri-urban or rural families, who need their personal vehicle to get around.

– "The electric car uses nuclear energy to function": this argument is linked to the historical opposition of a certain environmentalist fringe to nuclear energy, not because of its low carbon emissions, but because of the potential risks of power plants and disposal of waste.

Whether these positive or negative narratives are relevant depends on the structuring of the discussion topics leading to the narration of positions for or against electric vehicles. The black sheep of the electric car is first and foremost its battery, as well as exploitation of the minerals and rare earths necessary for its manufacture. Topics related to price, energy mix, charging stations, range and vehicle performance are also discussed. In addition, there are also specificities depending on the social network; on Twitter or Facebook, opinions are rather negative about electric vehicles, while on YouTube and Reddit, engagement is rather positive. In the media, positions vary, and it is precisely the heterogeneity of the data in our corpus that renders this visible.

10.5. Conclusion

A divisive subject, the electric vehicle is one of the solutions responding to the climate emergency. On the contrary, this subject resonates with a set of themes that go far beyond the scope of the vehicle itself.

Indeed, our relationship to technological innovation, the logic of mineral extraction, public transport policy or even the role of cars in public space all constitute elements structuring the positive and negative narratives produced in the public media sphere.

Although the electric vehicle is one of the solutions allowing a response to ecological and environmental issues, the narratives in circulation show that its use must unavoidably be accompanied by concrete measures, so as not to forget important questions in terms of communication and public action, but also by a model for structuring public space and dependence on certain raw materials.

Like any new technological production based on mineral extraction, the consequences of the industrial requirements for electric vehicles remain disastrous in some countries – and these consequences are increasingly visible and discussed in the public sphere.

10.6. References

Alexander, R. (2009). *Framing Discourse on the Environment: A Critical Discourse Approach*. Routledge, London.

Audet, R. (2014). The double hermeneutic of sustainability transitions. *Environmental Innovation and Societal Transitions*, 11, 46–49. doi: 10.1016/j.eist.2014.02.001.

Baker, P. (2006). *Using Corpora in Discourse Analysis*. Continuum, London.

Biros, C. (2014). Les couleurs du discours environnemental. *Mots. Les langages du politique*, 105, 45–66. doi: 10.4000/mots.21688.

Brady, J. and O'Mahony, M. (2011). Travel to work in Dublin: The potential impacts of electric vehicles on climate change and urban air quality. *Transportation Research Part D: Transport and Environment*, 16(2), 188–193. doi: 10.1016/j.trd.2010.09.006.

Calef, D. and Goble, R. (2007). The allure of technology: How France and California promoted electric and hybrid vehicles to reduce urban air pollution. *Policy Sciences*, 40, 1–34. doi: 10.1007/s11077-006-9022-7.

Canbing, L., Cao, Y., Zhang, M., Wang, J., Liu, J., Shi, H., Geng, Y. (2015). Hidden benefits of electric vehicles for addressing climate change. *Scientific Reports*, 5, 9213. doi: 10.1038/srep09213.

Carlino, V. and Stein, M. (eds) (2019). *Les paroles militantes dans les controverses environnementales. Constructions, légitimations, limites*. Éditions de l'Universitaires de Lorraine, Nancy.

Castellano, S., Maâlaoui, A., Schrempf, J. (2011). Le développement durable comme mode de prévention des risques énergétiques : une approche par les capacités d'absorption. Le cas de la voiture électrique chez Renault. *Management & avenir*, 42(2), 359–376. doi: 10.3917/mav.042.0359.

Garric, N. (2012). Construire et maîtriser l'hétérogénéité par la variation des données, des corpus et des méthodes. *Langages*, 187(3), 73–92. doi: 10.3917/lang.187.0073.

Kopytowska, M. (2015). Mediating identity, ideology and values in the public sphere: Towards a new model of (constructed) social reality. *Lodz Papers in Pragmatics*, 11(2), 133–156. doi: 10.1515/lpp-2015-0008.

Maillat, D. (2013). Constraining context selection: On the pragmatic inevitability of manipulation. *Journal of Pragmatics*, 59(2), 190–199. doi: 10.1016/j.pragma.2013.07.009.

Mougenot, B. (2015). Automobile et nouveaux modèles économiques de la mobilité électrique, au cœur d'une diversité institutionnelle. *Innovations*, 46(1), 71–88. doi: 10.3917/inno.046.0071.

Nanaki, E. and Koroneos, C. (2016). Climate change mitigation and deployment of electric vehicles in urban areas. *Renewable Energy*, 99, 1153–1160. doi: 10.1016/j.renene.2016.08.006.

Nykvist, B. and Nilsson, M. (2015). Rapidly falling costs of battery packs for electric vehicles. *Nature Climate Change*, 5, 329–332. doi: 10.1038/nclimate2564.

Poupon, L., Philipps-Bertin, C., Bobillier Chaumon, M.-E., Kalampalikis, N. (2017). L'acceptation de la voiture électrique ; un agent transformateur des pratiques socio-domestiques. *Activités*, 14(2). doi: 10.4000/activites.3044 [Accessed 3 January 2023].

Wagener, A. (2019). *Discours et système. Théorie systémique du discours et analyse des représentations*. Peter Lang, Brussels.

Wagener, A. (2020). Hypernarrativity, storytelling, and the relativity of truth: Digital semiotics of communication and interaction. *Postdigital Science and Education*, 2(1), 147–169. doi: 10.1007/s42438-019-00066-7.

Wagener, A., Hourcade, R., Le Bart, C. (2021). Discours climatosceptiques. *Mots. Les langages du politique*, 127. doi: 10.4000/mots.28695 [Accessed 3 January 2023].

11

What Has Ecological Transition Become the Name For?

The publication of the IPCC report on August 9, 2021, once again helped place climate change on the French media agenda, despite its unfortunate competition with the highly anticipated arrival of Argentine footballer Lionel Messi at the Paris-Saint-Germain club. Although the fight against climate change is now a major issue of concern, part of "more or less shared cultural frames of reference" (Campion 2016, p. 59), it is also part of a narrative of the Anthropocene which is more vast and difficult to delimit and dependent on numerous discursive formulae, including that of ecological transition.

The success of this expression, to which the "transition towns" movement (Hopkins 2010) has widely contributed, its circulation through bottom-up or multi-level social dynamics (Laigle 2014) linked to rhizomatic practices of eco-citizenship and multi-stakeholder governance (*ibid.*), its territorial dimension, the controversies with which it is associated (Krauz 2014), and its placing on the political and media agenda (Oudot and de l'Estoile 2020) further widen its scope. The socio-discursive trajectory of the ecological transition thus follows the recommendation *Let it go where it wants to go*, advocated by Rob Hopkins (2010). If the transversality and elasticity of this notion are a vector for enriching the narrative and practice of environmental issues, they nevertheless raise challenges for discourse analysis when it comes to understanding the more or less implicit sources of their sedimentation within common sense.

Numerous works have contributed to highlighting the media's ability to affect the perception and formulation of social or political objects (Page et al. 1987; Gamson and Modigliani 1989; McCombs 2005) including environmental issues

Chapter written by Brigitte SEBBAH and Natacha SOUILLARD.

(Comby 2015, 2009), with journalistic coverage producing "information" effects (Gerstlé 2001) and framing effects (Baylor 1996). Conversely, the latter is not impervious to socio-political dynamics, and media coverage is thus part of the co-construction of public problems, the narrative around them and their visibility (Neveu 2015).

Thus, and in order to contribute to the understanding of the narrative around environmental issues, this chapter seeks to "unfold" what the presence on the media agenda of the ecological transition leads to in terms of framing, and the extent to which analysis of these frameworks may shed light on the processes of requalification of the issues related to it. In other words, it is a question of focusing on the semantic scope of this elastic notion. We thus make the double hypothesis that not all the environmental issues involved in the notion of ecological transition are subject to the same framing, and that these framings are particularly sensitive to political prioritizations, "territorial receptions" (Rudolf 2012) and to phenomena of politicization or depoliticization (Comby 2015).

To investigate these questions, we chose to conduct a contrastive study of the dominant or at least significant media framings of the notion of ecological transition, in two types of mainstream press in France, the national daily press (NDP) and regional daily press (RDP)[1], in order to identify possible differences in focus at the thematic, discursive and territorial levels under consideration.

We compiled a corpus composed of 12,330 articles, extracted from the Europresse database and containing the expression "ecological transition" at least once. It should be noted here that these are mainly articles from the paper versions of newspapers with, at the margins, articles that have been republished on the Web. In order to study possible evolutions in media coverage, the chronological range extends from August 1, 2018, to October 1, 2020. The methodology is based on textometric analyses with divisive hierarchical clustering implemented in the IRaMuTeQ software (Ratinaud 2014). We used divisive hierarchical clustering (DHC) which forms lexical groupings that can be interpreted as "stabilized lexical worlds" (Reinert 2008) which may be interpreted in context due to the preserved metadata[2].

---

1 We chose to conduct a contrastive reading of the French RDP and NDP, based on publications with different editorial lines and territorial bases. Thus, for the NDP, we collected articles from the following media: *Le Monde*, *Le Figaro*, *L'Humanité*, *Les Échos*, *La Correspondance Économique*, *La Tribune*, *Libération*, *L'Agefi Quotidien*, *La Croix*, *Aujourd'hui en France* (7,419 articles). Concerning the RDP, the articles collected come from media with a geographical base in the Occitania region: *La Dépêche du Midi*, *Midi Libre*, *L'indépendan*t, *Centre Presse Aveyron*, *La Nouvelle République des Pyrénées* (4,911 articles).

2 Sources, type of press, dates, article identifiers.

In three stages, our analysis aims to question the stability and circulation of the notion over the whole corpus, then to grasp what this notion covers in terms of framing and meaning, before questioning the articulation of this scope of meaning with the dimension of "doing" public and political action, which is here dominant in the corpus. How does this notion fit into media coverage? What do the media say about the ecological transition? What are the applications of this project concept?

11.1. Between the groundswell and the economic situation, the ecological transition buffeted but central

A first quantitative look reveals some of the characteristics of the media circulation of the ecological transition and avoids the pitfall of over-investing in the meaning of this notion. Indeed, we can already emphasize that this is a notion which circulates extensively over the period studied (143,785 occurrences), and slightly more than the notion of "sustainable development" (134,541 occurrences). It is then used as an interpretive signpost for numerous subjects. Finally, note its strong indexation to the global informational context. This significant media presence nevertheless shows some variations. The largest publication peak is in fact observable in February and March 2019, with more than 800 articles published, while the lowest point was reached in April 2020, during the first lockdown, with only 176 articles published.

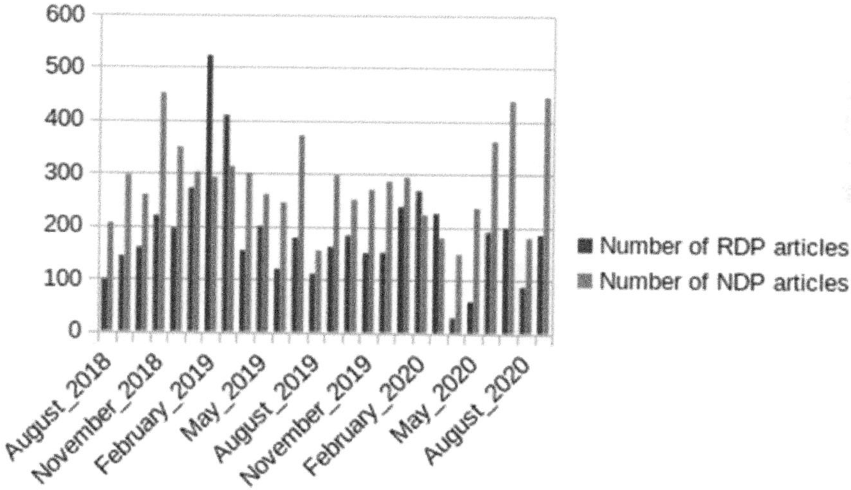

Figure 11.1. *Number of articles per month and by type of press. For a color version of this figure, see www.iste.co.uk/catellani/transitions.zip*

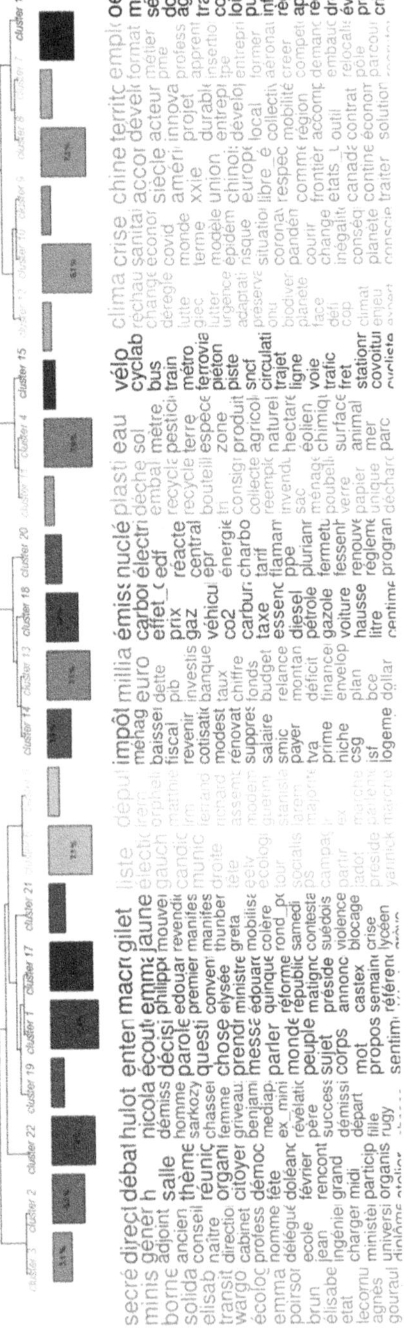

Figure 11.2. *Descending hierarchical classification (IRaMuTeQ) – 22 terminal clusters (97.42% of the classified corpus). For a color version of this figure, see www.iste.co.uk/catellani/transitions.zip*

From a comparative point of view, note also that more articles were published by the NDP with a shift in February and March 2019, during which period it was the RDP in Occitania which published more articles, thanks to the implementation of a great national debate, largely supported by local authorities. The gap between the RDP and NDP also widens from April to September 2020 in connection with the municipal elections, as well as with the fact that these two types of press do not necessarily follow the same agenda in terms of ecological transition and do not use the same modes of framing and prioritization of information on this subject.

The global panorama of the "lexical world" (Figure 11.2) reinforces this first characteristic, the 22 clusters testifying to this elasticity and a specific distribution of the notion. Also, two sets stand out from the dendrogram and seem to indicate not a single ecological transition, but ecological transitions.

Indeed, the first set, which includes clusters 3 to 5, deals with political and social news. In contrast, the second, ranging from cluster 14 to 16, relates to understanding and managing transitions. More diverse than the first, it is also more thematic and is subdivided into two subsets: clusters 14, 13, 18 and 20 reflect economic considerations, strongly linked to the energy aspect of the transition. Clusters 11, 4, 15, 12, 10, 9 as well as clusters 8, 7 and 16 relate more to the challenges of ecological transition (risk management, pollution, warming, crises) linked to territorialization (spatial development and use, territorial economic development).

An analysis of the chronological distribution of these "lexical worlds" reveals a relatively stable notion, with no clusters completely disappearing over the period studied. However, certain clusters are more limited in time, proof of a strong indexation of the notion to the news cycle. A retrospective calendar of the ecological transition thus emerges quite clearly through various events allowing measurement of the large volume of episodic framings in comparison to more thematic and transversal framings of the notion.

Media coverage of the ecological transition thus depends on the adoption of laws and international agreements[3]. Next, the period studied was marked by an important

3 "Agriculture and food law" (November 1, 2018); "energy-climate" law (November 8, 2019); finance law (December 28, 2018); law relating to the fight against waste and the circular economy (February 2, 2020); JEFTA (signing of the treaty at European level on February 1, 2019); ratification of CETA (July 23, 2019). As to CETA, this finance law particularly relates to the creation of two tax instruments, the TICPE (domestic tax on the consumption of energy products, otherwise called the "carbon tax") and the CITE (energy transition tax credit).

electoral calendar[4]. Then, an institutional agenda had an impact on coverage, with the publication of several IPCC Special Reports[5], including the one on land use and the cryosphere (August and September 2019). Added to this is the social agenda with the social movements and the political response to them which marked the period: the Yellow Vests from 2018, followed by the great national debate in 2019[6], and the climate marches. Finally, it is worth emphasizing the impact of industrial disasters such as the Lubrizol fire (September 2019), natural disasters (floods, droughts) and finally, the health crisis. The media coverage of the ecological transition therefore appears to be part of a double dynamic of visibility: dependent on a multidimensional event, the stabilized notional scope of the notion is, furthermore, clearly dominated by expert, economic and managerial discourses in the NDP and RDP overall.

11.2. Ecological transition: a depoliticized expert notion

In 2009, Comby identified three forms of depoliticization of the climate issue: individual empowerment rather than a collective approach, equalization of the positions of the actors in relation to the issues, and finally bringing the question into everyday life (Comby 2009). Our observations add three dimensions to these analyses: a disembodied narrative of the climate, a disconnect between certain major environmental issues and operationalization of the ecological transition as public policy, and finally, a dominant economic and managerial narrative.

Although, as Comby points out, journalists tend to counter the climate issue's lack of proximity, its excessive abstraction, by resorting to "bringing it into everyday life", our results instead reveal a very compartmentalized alternation between generalizing it and particularizing it. In RDP, the anecdotal dimension is more supported than in NDP, and relates to a framing that may be described as "episodic" (Lyengar 1991). It has the effect of evacuating the complexity of the concept and of placing itself within the daily experience of the operationalization of public policy, of situated and non-global issues.

Also, and less surprisingly, in NDP, few links are made between global warming and examples located in specific or hyperlocal territories. The framing is mainly thematic, focusing on major issues. Furthermore, it seems to abstract from the lived experience of climate, whether expressed by evocation of specific territorial situations

4 European elections (23/26 May 2019); French municipal elections (15 March/28 June 2020).

5 Intergovernmental Panel on Climate Change.

6 The great national debate was a national dialogue set up by the government in response to the Yellow Vests social movement. The stated themes included that of the ecological transition.

or testimonial narratives. In RDP, on the other hand, the weather (inclement weather and natural disasters) tends to replace the climate. However, we note, in NDP as well as in RDP, that the relatively rare traces of orality in the corpus are not associated with the reported words of "ordinary" citizens, but mostly with those of political leaders and experts (cluster 1). This framing is common to these two types of press, testifying to the frequent use by journalists of institutional and/or political sources.

Finally, the prevalence of economic discourse, with a strong technical dimension, in RDP and NDP, associated with the energy aspect of the transition (clusters 14, 13, 18, 20), tends to deprive the notion of ecological transition of its force of conflict or polarization, thus contributing in a way to its depoliticization (Cussó and Gobin 2008; Gourgues 2018), while once again testifying to the hegemonic character of economic discourse (Bonnet and Canu 2017). The ecological transition thus appears mainly sometimes as a policy to be financed by means of fiscal and financial "instruments" (Halpern and Le Galès 2011)[7], sometimes as a lever for economic recovery at the French and European level. As such, the post-Covid-19 recovery plan of the summer of 2020 for "a greener, more digital and more resilient Europe" is the subject of extensive coverage (cluster 13).

> In addition, the government plans to take advantage of low rates to borrow up to 50 billion euros on the markets in order to set up an investment fund intended to finance the ecological and technological transition to support education and to build up financial reserves in the event of a shock (*Le Monde*, October 11, 2019 (cluster 13)).

Discontinuity between the general and the particular, absence of lived experience of the climate, pervasiveness of an economic vision; the ecological transition seems to refer more to a potential project, even one in progress, rather than an existing reality, largely unthought of, in the ecology. The modalities of the "adaptationist" operationalization (Gourgues 2018) of this nascent public policy emphasize the dimensions of planning and management, with, however, some traces of conflict and tension.

11.3. Implementing the ecological transition, between territorial economic development and a changing world

Like the great national debate, which was largely operationalized by local authorities and almost exclusively discussed by the RDP (cluster 22), the

7 While the Yellow Vests movement criticizes one fiscal instrument for ecological transition, the media coverage tends to favor an educational approach to measures or to discuss the impact of this tax on household income (cluster 14).

"implementation" of the ecological transition operates in a territorialized manner in RDP or NDP with the central themes of employment and territorial economic development (clusters 8, 7, 16). Public action is itself strongly mobilized (cluster 8) from the "project" angle following on from related measures (local, regional, rural development contracts, ecological transition contracts, etc.). This decentralized, project-based planning is also reflected in the frequent evocation of multi-stakeholder territorial governance for which the ecological transition is presented as a lever of activity. It is also significantly associated with technological innovation, as attested by terms like "start-up", "cleantech", "energy", "sustainable" and "renewable". Finally, this multi-stakeholder governance itself appears as an issue and not as an achievement, through questioning and specific mentions of different types of stakeholders (citizens, businesses, associations, politicians, etc.), which testifies to this base's lack of stability in a sense.

> How to revive and revitalize local economies? Building territorial projects with all local stakeholders, developing a local development contract with the participation of all stakeholders such as citizens, businesses, and the associative sector, and ensuring that the associated resources are ensured (*La Tribune France*, September 6, 2019 (cluster 8)).

> Its ambition is to demonstrate through action that ecology is a driver of the economy by developing local employment through ecological transition, structuring sectors, developing new training and acting with all territorial stakeholders, both public and private, to translate the ecological transition into concrete action (*La Dépêche du Midi*, January 16, 2020 (cluster 8)).

> The main objectives of this contract are to prove that ecology is an economic engine for local development and that the ecological transition is the business of all stakeholders in the territory, both public and private, in a partnership approach (*Midi Libre*, October 3, 2019 (cluster 8)).

> Concluded for a period of 4 years, the rural contract allows co-ordination of public action in the territories and support for projects relating to accessibility to services and care, development of attractiveness [...] (*La Dépêche du Midi*, July 18, 2020 (cluster 8)).

This projective vision of public action is to be associated with a managerial categorization linked to spatial development that is clearly perceptible in clusters 4 and 15: "zone", "site", "natural space"; "urban", "municipality", "mobility", etc., in connection with the installation of equipment for water reprocessing, the production of "green" (principally wind) energy, the artificialization of soils, industrial pollution and infrastructures for active modes of transport.

> [...] including around twenty in the Occitanie region, notably electromechanics working at sea, should thus be created during the operational phase of the pilot wind farm. The project will consolidate the offshore wind industry by promoting local and regional know-how, it is announced (*L'Indépendant*, December 3, 2018 (cluster 4)).

> At its level, the department is leading the territory's ecological transition, particularly in relation to transport with the development of 25 new carpooling areas, and we are going to finance the creation of five fast and secure express bicycle networks in the Toulouse metropolitan area by 2023 (*La Dépêche du Midi*, January 17, 2020 (cluster 15)).

> We must prohibit any artificialization of land if restorations of commercial, artisanal or industrial brownfield sites are possible within the existing urban envelope, and firmly and definitively protect peri-urban natural and agricultural spaces (*Libération*, July 30, 2020 (cluster 4)).

A more in-depth exploration of cluster 8, however, reveals less common narratives, predominantly more social and environmental ("solidarity", "culture", "environment", "democracy", "quality of life", "proximity", etc.). This is a register associated with the electoral programs of the municipal elections of June 2020 and which embodies an ideal, that of local living together. It is also with this cluster that the term "sustainable" is most associated, reflecting the sustainable development objectives of the territories. At the level of the RDP, if the ambitions of the region with regard to "positive" energy are described as a vector of economic development, a "new world opportunity", they are also questioned as to their compatibility with the "preservation of landscapes", "biodiversity", "attractiveness" as well as the "life of our rural territories". One of the narratives of this balance describes a trade-off summarized in terms of "end of the month, end of the world" and "solidarity-based development of territories", considering the "needs of the population".

> [...] invest in renewable energies that are accepted by the population, efficient, compatible with the preservation of landscapes and respect for biodiversity, and compatible with the attractiveness of rural territories (*Centre Presse Aveyron*, December 20, 2018 (cluster 8)).

> Armand Rivière, in his presentation, highlighted the priorities of the mandate, confirming his commitments to democratic life, the ecological transition, economic attractiveness, popular knowledge, solidarity and proximity (*Midi Libre*, September 3, 2022 (cluster 8)).

> We wish to support our fellow citizens in adapting to all these changes while applying the principle of solidarity and equality. We must also think about preserving the quality of life, our desire is to make it possible to live in a setting that offers protection in daily life and for the future (*La Dépêche du Midi*, February 19, 2020 (cluster 8)).

At a more general level and in the NDP, other tensions are identifiable around the terms "crisis" and "change", particularly represented in cluster 10, the most extensive in the corpus. Above all, it demonstrates the way in which the health crisis has overshadowed all media coverage and absorbed even the ecological transition. This is also where we see the appearance of a previously absent link between the health crisis and the climate crisis. This link is, however, framed more from the angle of the measures planned to revive "green" growth than from the angle of a possible correlation between the destruction of natural environments and the emergence of zoonoses. "Change", meanwhile, is discussed from the angle of the economic system which "deepens inequalities", "destroys ecosystems", and from the angle of lifestyles and consumption which must aim for a certain sobriety. The everyday implementation of this change is quite clear since the stories are often written at the level of the city, the local, the living environment. A balance emerges here between the actions of citizens and those of public policies for meeting the "challenge" of "economic development" of territories through ecological transition. It is also in this context that the anchoring of the transition in citizens' initiatives is more readily discussed.

> Cities in transition or ZADs [zones occupied by protesters] are all ways of producing and experimenting with other ways of life and consumption, other relationships between citizens (*L'Humanité*, March 15, 2019 (cluster 10)).

> […] They are at the heart of the new model of growth combining social responsibility, ecological transition and digital revolution, it is urgent to recognize this and to bring all of these subjects to the level of a global political reflection (*Les Échos*, July 6, 2020 (cluster 10)).

> […] Furthermore, 81% of French people surveyed by Elabe say they are obliged to change their habits and adopt a more sober lifestyle. Again, this is no longer an abstract subject, but a lived reality, insists Laurence Bedeau (*La Tribune*, November 15, 2019 (cluster 10)).

Finally, and more specifically in the PQR, "risks" also reflect an ambivalence, between alarmism, even catastrophism, management and associated controversies. The most recurrent topics, presented from the point of view of their impact on the environment and on health, are mainly pesticides, air and water quality, overfishing, management of plastic and chemical wastes, food and industrial risks (clusters 4 and 11). Some subjects and risks are framed in a hyper-localized way, depending on controversies, such as bear attacks and agropastoralism or "5G" antennas.

> Paris and four other large cities yesterday banned the use of chemical phytosanitary products on their territory joining the anti-pesticide battle launched by rural mayors in the hope of putting pressure on the government which had mocked it as a publicity stunt [...] (*La Dépêche du Midi*, September 13, 2019 (cluster 4)).

> I reaffirm that the presence of bear populations is incompatible with the practice of agropastoralism, especially since the introduced species has a significantly more carnivorous diet than the authentic native species [...] (*La Dépêche du Midi*, September 29, 2018 (cluster 4)).

11.4. Conclusion

The scope of the ecological transition appears to be that of an intersectoral public policy, still fragmented and in the process of being implemented, whose normative horizons are rarely discussed, but which are operationalized. The results of this study firstly attest to a certain sensitivity of coverage of the ecological transition to the social and political situation. The "stabilized lexical worlds" then sketch an overview of the notion dominated by economic and managerial registers, between episodic and thematic coverage thus underlining a gap between a vision planned from "above" and an implementation of the notion in daily life with hyper-local focal points.

In short, it is the lived experience of climate which appears to be a blind spot in these two types of framing and which raises the question of the articulation between the "saying" of the ecological transition and its "doing". As to the question of "doing", in fact, we note the strong presence of an imperative for territorial economic development, structured by multi-stakeholder projects in which the citizen has little place. Conversely, the dimension of citizen initiative is little embodied, or even little narrated, tracing here a framing which in general tends to invisibilize not only the lived experience of climate but also the multiple controversies and polarizations to leave room for imperative and alarmist registers with an economic and managerial focus. The information content studied also demonstrates in a transversal manner a hesitation which seems common to all types of printed press around the notion of ecological transition, a hesitation which nevertheless reveals two types of approach in the treatment of ecological questions, one mobilizing political, economic or even social focal points, but in the service of ecological questions. The other approach tends to prioritize the economic and political question, thereby making ecological questions invisible, diluted in questions that transcend them. This hesitation seems to testify to the actualization, or persistence, in the narrative of ecological transition of a phenomenon of competition rather than real convergence between the paradigms of weak sustainability and strong sustainability.

11.5. References

Baylor, T. (1996). Media framing of movement protest: The case of American Indian protest. *The Social Science Journal*, 33(3), 241–255. doi: 10.1016/S0362-3319(96)90021-X.

Bonnet, V. and Canu, R. (2017). L'ordre des discours de la crise : un agencement hétérogène. *Mots. Les langages du politique*. doi: 10.4000/mots.22923.

Campion, B. (2016). Alarmisme et catastrophisme ? Rhétoriques de la peur dans les échanges en ligne relatifs aux changements climatiques. In *La communication environnementale*, Libaert, T. (ed.). CNRS Éditions, Paris.

Comby, J. (2009). Quand l'environnement devient "médiatique" : conditions et effets de l'institutionnalisation d'une spécialité journalistique. *Réseaux*, 157/158(5/6), 157–190. doi: 10.3917/res.157.0157.

Comby, J. (2015). *La question climatique : genèse et dépolitisation d'un problème public*. Raisons d'Agir, Paris.

Cussó, R. and Gobin, C. (2008). Du discours politique au discours expert : le changement politique mis hors débat ? *Mots. Les langages du politique*, 88, 5–11. doi: 10.4000/mots.14203.

Gamson, W.A. and Modigliani, A. (1989). Media discourse and public opinion on nuclear power: A constructionist approach. *American Journal of Sociology*, 95, 1–37.

Gerstlé, J. (2001). *Les effets d'information en politique*. L'Harmattan, Paris.

Gourgues, G. (2018). Participation : trajectoire d'une dépolitisation. *Revue projet*, 363, 21–28. doi: 10.3917/pro.363.0021.

Halpern, C. and Le Galès, P. (2011). Pas d'action publique autonome sans instruments propres : analyse comparée et longitudinale des politiques environnementales et urbaines de l'Union européenne. *Revue française de science politique*, 61, 51–78. doi: 10.3917/rfsp.611.0051.

Hopkins, R. (2010). *Manuel de transition. De la dépendance au pétrole à la résilience locale*. Écosociété, Montreal.

Krauz, A. (2014). Les villes en transition, l'ambition d'une alternative urbaine. *Métropolitiques* [Online]. Available at: https://www.metropolitiques.eu/Les-villes-en-transition-l.html.

Laigle, L. (2014). Une mise en mouvement de la transition écologique par la société civile ? Approches, enjeux et perspectives. *Colloque international "Les chemins politiques de la transition écologique"*, Lyon.

Lyengar, S. (1991). *Is Anyone Responsible? How Television Frames Political Issues*. University of Chicago Press, Chicago.

McCombs, M. (2005). A look at agenda-setting: Past, present and future. *Journalism Studies*, 6(4), 543–557. doi: 10.1080/14616700500250438.

Neveu, É. (2015). *Sociologie politique des problèmes publics*. Armand Colin, Paris.

Oudot, J. and de l'Estoile, É. (2020). La transition écologique, de Rob Hopkins au ministère. *Regards croisés sur l'économie*, 26, 14–19. doi: 10.3917/rce.026.0014.

Page, B.I., Shapiro, R.Y., Dempsey, G.R. (1987). What moves public opinion? *American Political Science Review*, 81(1), 23–43. doi: 10.2307/1960777.

Ratinaud, P. (2014). IRAMUTEQ, Interface de R pour les Analyses Multidimensionnelles de Textes et de Questionnaires. Un logiciel libre construit avec des logiciels libres [Online]. Available at: http://www.iramuteq.org.

Reinert, M. (2008). Mondes lexicaux stabilisés et analyse statistique de discours. *Actes des 9e Journées internationales d'Analyse statistique des Données Textuelles (JADT). Lexicométrica* [Online]. Available at: http://lexicometrica.univ-paris3.fr/jadt/jadt2008/pdf/reinert.pdf.

Rudolf, F. (2012). La réception territoriale du changement climatique ou comment le changement climatique contribue à l'émergence de territoires et de politiques climatiques spécifiques. *VertigO – La revue électronique en sciences de l'environnement*, 12. doi: 10.4000/vertigo.11825.

12

Labels as a Process of Institutionalization of Environmental Themes

The ecological political record of recent years has been judged harshly by the media and the courts, in France as in other countries. However, since the spring of 2021, environmental issues have successfully become part of the European and French institutional agenda via major initiatives and funding. The "climate action plan" was launched with a desire for green growth projects, ecological transition and renewable energy. Impacting an increasing number of areas (automobiles, food, agriculture, technologies, etc.), this plan is part of a vast movement of institutionalizing environmental themes, notably through the national climate strategy, Agenda 21, and today the Sustainable Development Goals and Agenda 2030.

The concept of ecological transition appeared as peripheral and alternative a few years ago among urban and permaculture communities. It appears particularly as a *bottom-up* concept, which became institutionalized initially based on the concept of energy transition (factor 4, transition towns) and then developed into ecological transition, a much broader concept. It is based on a "green budget", as embodied in the 2022 finance bill, supplemented by "green bond issues" (government bonds), by a "future investment program", by a "plan for the ecological transition of VSEs and SMEs" and so on. A shift could therefore potentially be at work towards taking ecological issues into consideration in all public policies in France. This global movement is supported by an allocation of funds, as well as by new norms, and consequent supporting discourses, leading to changes operating at the level of construction of norms. For example, negotiations in business circles lead company stakeholders to writing and then to communicating around labels, corresponding to processes of reconfiguring places of negotiation and spaces of speech.

Chapter written by Catherine LONEUX and Céline PASCUAL ESPUNY.

At a methodological level, to analyze this institutionalization of labels, we are included in a current of information-communication sciences which has long been interested in charters, labels, benchmarks (Huët 2008; Pascual Espuny 2008; de la Broise 2013; Loneux 2016), and considers it possible to partially capture the "social dynamics of meaning production" (Huët 2008) implicit in the discourses and ethical stories of management. Likewise, these labels are seen in this current of research as "corpora of documents" that are useful to professionals in organizations, whose "segmentation operates not only upon the mode of differentiating territories with variable geometries and promoting their sovereignty, but also through work for the support and supervision of practicing professionals" (de la Broise 2013, p. 33).

It is through the prism of this dynamic and communicative reading of the normative that we will proceed to interpret our two area of research concerning the *France relance* ("France recovery") plan and the "envol" (taking off) label. We choose to analyze them particularly using the CAO ("communication approaches to organizations") frame of analysis (Bourdin et al. 2007, 2008), to characterize the rise of soft law. This frame aims to isolate some observables via three organizational inputs:

– micro, the pragmatic level of work situations and interactions that are the location of knowledge that is informal and on-the-job (professional equality, ecology, occupational health, etc.) but whose value is comparable to that of scholarly knowledge for the stakeholders;

– meso, the level of organizational production processes, which provide the opportunity to connect people with each other. The coordination of stakeholders, common frameworks and professional credos are shared references, which thus stabilize over time and are worth observing;

– macro, the political level corresponding to strategy and discourse around the purpose of the company, around its social utility, to the languages of "corporate social responsibility", which responds to the pressure of societal expectations and helps manage reputational risk.

At a scientific level, two research questions emerge around this problem of the rise of new assemblages of stakeholders and norms, which lead to changes in normative construction. These include:

– connections between regulation and communication;

– connections between regulation and changes in society.

12.1. Regulation and communication

The sociology of law can be used to understand the link that operates between the construction of a law and the context of communication. Here, we refer to the "texture of the law" (Thibierge 2003) to evoke the reciprocal influences acting upon the law. Paul Amselek (1982) similarly evokes "the indecisive content of the law".

The law is finally said to be the formalization of citizen expectations, managerial and institutional discourses of ethics and the professional context which embodies "good practices". There are therefore two key concepts to be considered, that of ethics and that of soft law, which constantly interact. Soft laws are proposed and added to the rule of law in a sort of question/answer/adjustment sequence. In a way, the law follows on from ethics. "Problems" are "discussed" publicly or in companies, and finally, the law emerges, either to rectify poor professional practices or simply to regulate them. Thus, companies "produce", in a symmetrical fashion to laws which themselves arise from state and legal powers, professional recommendations arising from a soft law approach.

Our theoretical info-communication framework therefore intersects with notions of the sociology of law. Indeed, this theoretical current of law (Amselek 1982, 1991; Thibierge 2003, 2012) is in a way part of a communicational approach centered on the interactions of stakeholders and the evolution of legal technique seen through the prism of its phenomenological dimension. For example, it aims to focus its work, not on legal texts, but more on legal phenomena, their relationship to the collective conscience, and to the stakeholders who design the laws.

In this perspective, we are particularly interested in the labile and plastic dimension of the construction of a norm (Loneux 2016). Thus, "hard" law is based on the rules of positive law, while "soft" law is based on flexible and evolving standards. This flexibility allows companies to experiment with different systems of "governance" depending on their market, their shareholder base, their size and their era. It makes it possible to apply principles from which societies are supposed to not derogate to a multitude of situations.

Concretely, soft law relies on three levers to progress: first, it takes root in rather vague organizational processes in which the interplay of players is not predetermined or framed by too much immediately apparent hierarchy. Secondarily, the promoters of soft law control the space of interrelation in which it develops. Finally, and we will focus on this lever: media coverage of these ethical measures is strong (Pascual Espuny 2014).

Companies publicize these accreditation initiatives. The comply or explain approach attests to this ongoing communication around the ethical norm, as a condition for being imposed (Circle des Affaires 2020). In this regard, we may mention the role played by compliance officers who communicate in two stages. Phase 1 consists of *building a shared ethical culture*, of responsibility and integrity. To do this, communicators seek to identify ethical risks and establish a common language within the company. Phase 2 involves *highlighting these labels in order to contribute to performance*. In this regard, beyond publicizing the CSR indicators that are to be respected, it is up to the compliance officer to raise awareness of the economic opportunities represented by a successful ethical approach: reduced criminal risk, less costly investor financing, reputation and employer brand protection.

This media coverage in the business environment reveals an interest in adapting to the context of citizen and regulatory pressure (preservation of the environment, food, etc.). It is noted that these labels are an easy way to adjust to such demands due to their flexibility of use. They do not have the force of law, and meet the expectations of the general public and investors. Structured actions and discourses that legitimize and refer to one other surround these ethical norms. This makes it possible to set up unitary representations around values, ideals and convictions, which are defended at the environmental level and governed by these self-regulations.

Texts, writings and words with an ecological, economic or societal vocation "circulate" within and in the environments of organizations, and are associated with long-lasting socialized actions (Bernard 2010), driving logics of change in groups. In this sense, the dynamics of CSR become prioritized in public debate; defense of biological diversity and the values of nature are integrated into the corporate decision-making process.

12.2. Regulation and societal change

Regulations of the soft law type also merit scrutiny from a more institutional angle: our society is changing in terms of regulation. Our objective at this stage of our discussion is to try to understand socio-economic, regulatory and institutional changes in society. A form of hybridization between market and social norms exists (Loneux and Huët 2009), and the notions of institution and border seem heuristic in this regard, in connection with these changes.

Associated with the notion of institution, changes in representations, mentalities and imaginaries of organizational functioning are emerging. Content effects (ideological, symbolic or semiotic) intervene and establish the regulatory form of soft law and thereby, the organizations that may enact them. In terms of theories of regulation, the relationship between discourse, action and institutionalization must be emphasized. The circulating languages of ethics have an impact on knowledge and cultures.

The boundaries between professional, public and private spheres are interwoven. Thus, a change in guidance is taking place in business: stakeholders, not always well defined, play a strong role. Associations, third places, and local actors are organizing for ecology and solar power, for example.

We then see a "de-statization" of the construction of law, and self-regulation aims to replace the classic, standardized and collective forms of collective negotiation of rules with a privatized form of regulation. "Traditional" law recedes under the influence of a regulation made up of relationships of hybridity, of co-construction. The positive theory of law is in tension with the intensifying contamination of self-regulatory norms. Isabelle Duplessis speaks of the "vertigo" of soft law (Duplessis 2007).

To capture this connection between communication and regulation over these vast ecological movements, we have chosen to focus our analysis on two recent case studies, both linked to French public policies, but which represent different levels of action, forms and measures.

Our idea is to mirror these two areas, to enable us to propose an initial qualification of the link that currently exists between communication and regulation.

Our case studies provide organizations with possible accreditations that are entirely characteristic of this form of corporate institutionalization, as already mentioned: "differentiate yourself and give credibility to your commitment; attract new customers" (CCI France 2021).

Our first area, the label associated with the *France relance* plan, marks the achievement of national accreditation and funds obtained from the 100 billion euros of the *France relance* plan[1], of which 30 billion will be entirely devoted to the ecological transition. Its announced objective is to make ecology the main lever for the recovery and transformation of the French economy. Here we will question, through organizational communications, a top-down and macroeconomic movement which leads and frames a planned and desired transition.

1 See: https://www.economie.gouv.fr/plan-de-relance.

Figure 12.1. *Ministry of the Economy, Finance and Industrial and Digital Sovereignty: presentation of the deployment of the France relance plan*[2]*. For a color version of this figure, see www.iste.co.uk/catellani/transitions.zip*

We have chosen the second area, the "envol" label, as an "at your service" measure, which must be grasped and used at stakeholder level, through a declared institutional desire of the French Chamber of Commerce and Industry (CCI France) and of Bureau Veritas, and falling within the framework of the PACTE law of 2019[3]. The "envol" eco-label[4] requires a voluntary commitment by the company to the environment; it allows companies to be part of a "community" called "Lucie" and to participate in events, to benefit from "shared communication or even tools to structure their commitment" (CCI France 2021).

2 See: https://www.economie.gouv.fr/plan-de-relance#.

3 The Action Plan for Business Growth and Transformation (PACTE) promulgated on 22 May 2019 "aims to remove obstacles to the growth of businesses at all stages of their development: from their creation to their transmission, through their financing. [...] It also allows companies to take better consideration of social and environmental issues in their strategy."

4 See: https://www.envol-entreprise.fr/.

In both cases, we mobilize an analysis of discourse and interactions to understand the issues in terms of comprehending societal changes, and an analysis of semiotic and linguistic inspiration in order to grasp the "democratic procedural model".

Figure 12.2. *Presentation of the envol approach*[5]*. For a color version of this figure, see www.iste.co.uk/catellani/transitions.zip*

The semiotic-inspired analysis thus clearly shows the *France relance* plan as an outlet for national power. Syntactic and linguistic choices are dominated by numbers

5 See: https://www.envol-entreprise.fr/.

and indicators, promoting an expert and technocratic approach. A vocabulary of action and reconquest accompanies this semantic universe of power and control. Semiotic choices reinforce this desire for seriousness: power is embodied in the images and photographs, often taken from a low angle, in the precision of the lines and the respect for alignment and symmetry in the construction of means of communication. Seriousness is embodied in the neutral color codes, gray, white and dark green, in compliance with strong institutional codes (immediate presence of the logo, the French and European flags), in the strength and enhancement of the "*France relance*" stamp.

The creation of the norm thus appears voluntary and is directly linked to the institutional influence of the codes used. *France relance* also offers in intertext a discourse of acceptance of the anthropological mutations at work in its written media. We note the political dimension of its media productions: it is a question of convincing that ecology is one of the levers of the recovery and transformation of our economy. We detect "practice-prescribing watchwords" (Chabot 2015) around the communication "formulas" (Krieg-Planque 2010) which regularly emerge in discussions of sustainable development.

The *France envol* label invites us to another reading. Here, and despite support for the label by CCI France, also very institutional, the goal is the creation of a community. This translates into a refined semiotic construction, a search for space and a play on emptiness and purity, which gives envol's communication a strong modern and Zen connotation. Five colors are distinguished in the communication media (notably the website): the background is beige and neutral, while black and white are used in turn for the text blocks. Graphic and visual elements are marked by blue and green colors. This blue-green color chart is expressed in curved, round shapes, and also in a search for purity, especially in the diagrams on the website. Roundness is also found in the gaps in space, embodied in the "V" of the envol label. A semantic trilogy is established throughout the communication media, which we characterize around the key concepts of "concrete" (no formalism, simple and clear vocabulary), "accessible" (highlighting the affordable costs of the accreditation process, ease of reading and identification) and "adaptable" (local, territorial, support).

The envol label establishes a clearly identifiable universe for companies, and for us it is a form of institutionalization: for CCI France, the stated objective is to "differentiate itself and give credibility to its commitment: attract new customers". The Lucie community thus created gives visible form to a state of mind based on companies' mutual support and commitment.

These two digital spaces, showcases for an institutional desire to enact the ecological transition, show different intentions, as well as places and "facades", in

Erving Goffman's sense (Goffman 1973): the place given to stakeholders in making use of the site, in navigating it, as well as in the constructed semiological universe, and conversely, that of the institution, sheds light on the institution of the soft law form of regulation, here embodied in these two case studies. If *France relance* is an essentially informative "at your service" site, the relationship is established within a top-down and very institutional form of communication. The desire of the envol label is quite different, with the incentive for "partnership" and the provision of digital tools relying on the target audience making use of the measure. The success of this "at your service" depends on accessibility and more elaborate and above all more interactive use. The roles and places of each are therefore particularly representative of the changes in guidance enacted by the institutions and the roles available to stakeholders.

To conclude on these organizational and institutional changes associated with soft law, which we observe in the context of ecological transition, the salient points which concern us are therefore the following: increasing self-definition of society, redrawing of boundaries between societal actors, renewal of the democratic procedural model and full-scale implantation of companies in public debate.

The role given to stakeholders is becoming greater, which is currently establishing a trend towards naturalization of the corporate citizenship approach. A semiotic stabilization is also emerging around a transition that would adjust to a wider audience, with a choice of the concrete and of simplicity, an engagement and the emergence of a community of meaning. This stabilization is driven by a graphic universe that stabilizes and fully participates in the intention and in the level of participation offered to stakeholders in constructing the norm, as well as in the success of the devices enacted.

12.3. References

Amselek, P. (1982). L'évolution générale de la technique juridique dans les sociétés occidentales. *Revue du droit public et de science en France et à l'étranger*, 2, 275–294.

Amselek, P. (1991). La teneur indécise du droit. *Revue du droit public*, 107, 1212.

Amselek, P. (2012). *Cheminements philosophiques dans le monde du droit et des règles en général*. Armand Colin, Paris.

Bernard, F. (2010). Pratiques et problématiques de recherche et communication environnementale : explorer de nouvelles perspectives. *Communication & organisation*, 37, 79–89. doi: 10.4000/communicationorganisation.1275.

Bourdin, S., Bouillon, J.-L., Loneux, C. (2007). Migrations conceptuelles. Construction des approches communicationnelles des organisations dans les Sciences Humaines et Sociales. *Communication & organisation*, 31.

Bourdin, S., Bouillon, J.-L., Loneux, C. (eds) (2008). *Approches communicationnelles des organisations. Perspectives internationales*. Presses universitaires du Midi, Toulouse.

de la Broise, P. (2013). La lutte pour la reconnaissance ? Codes, chartes, référentiels et autres manifestes de professionnels français de la communication. *Communication et professionnalisation*, 1, 33–49 [Online]. Available at: https://ojs.uclouvain.be/index.php/comprof/article/view/233/183.

Chabot, P. (2015). *L'Âge des transitions*. Presses Universitaires de France, Paris.

Duplessis, I. (2007). Le vertige et la soft law : réactions doctrinales en droit international. *Revue québécoise de droit international*, 245–268.

Goffman, E. (1973). *La mise en scène de la vie quotidienne*. Éditions de Minuit, Paris.

Huët, R. (2008). Proposition méthodologique pour saisir les dynamiques sociales de production des textes. *Sciences de la société*, 78, 173–180.

Krieg-Planque, A. (2010). La formule "développement durable" : un opérateur de neutralisation de la conflictualité. *Langage et société*, 134(4), 5–29.

Loneux, C. (2016). La responsabilité sociale des entreprises comme *Soft Law* : formes et enjeux de régulation, de dialogue et de frontières. *Revue française des sciences de l'information et de la communication*, 9. doi: 10.4000/rfsic.2284.

Loneux, C. and Huët, R. (2009). Le tournant institutionnel de la RSE à l'épreuve de l'analyse communicationnelle. *Revue française de socio-économie*, 4(2), 117–134. doi: 10.3917/rfse.004.0117.

Pascual Espuny, C. (2008). Comment les organisations se saisissent-elles de l'image verte ? *Communications & organisation*, 34(2), 39–52. doi: 10.4000/communicationorganisation.572.

Pascual Espuny, C. (2014). La société civile, de l'alerte à la controverse médiatisée. *Communication & organisation*, 45, 115–126. doi: 10.4000/communicationorganisation.4531.

Thibierge, C. (2003). Le droit souple. Réflexion sur les textures du droit. *Revue trimestrielle de droit civil*, 599.

Professional documents

CCI France (2021). [Online]. Available at: https://www.cci.fr/actualites/le-label-envolENVOL.

Cercle des affaires (n.d.). [Online]. Available at: https://www.lecercledelacompliance.com/compliance-officers/.

France relance (n.d.). [Online]. Available at: https://www.economie.gouv.fr/plan-de-relance.

13

Comparative Analysis of Ecologist Discourses Produced for the Web in France

Ecology is a polysemous term, around which revolve "language materials and discourses, between acceptability, reappropriation and diversion" (Bonnet and Geslin 2019, p. 10). Various actors work to semanticize the term, trying to define what ecology should refer to. In this chapter, we will focus on the discourses that certain ecologist organizations use across digital media spheres. Considering that there exists a nebula of various ecologist organizations (Vrignon 2017), we will proceed to a selection of organizations representative of contemporary French ecologism, and then investigate by what modalities these might succeed in deploying an ecologist discourse over digital media arenas.

We will first distinguish which info-communicational channels might be used by environmental organizations, then we will investigate whether the discourses they deploy demonstrate a diversity of thematic treatment or if, on the contrary, they prove to be somewhat homogeneous. To do this, we will perform an analysis of their discourse using the IRaMuTeQ textometry software (Ratinaud 2014). It allows for a morpho-lexical analysis using the so-called Reinert classification method (Reinert 1983), which consists of identifying situations of co-occurrences between terms and their semantic proximity in order to distinguish the main themes that emerge from discourses. We will thus examine what types of claims may be defended by environmental organizations on the Web and through what info-communicational channels they may be deployed.

Chapter written by Jules DILÉ-TOUSTOU.

13.1. Ecology at the heart of a semantic struggle

Ecologism has its origins in various politico-scientific currents and could be considered a kind of naturalistic ideology, advocating the protection of natural ecosystems and living beings. This would lead to claims that we will consider to be ecologist arguments, aiming at a minimum "to exert influence by directing ways of seeing and thinking" (Amossy 2008, p. 1).

From this activist perspective, two original currents might be distinguished, that of environmentalism and political ecology. The first would consist of the existence or otherwise of certain environmental aspects in the management of public policies, while the second would aim to offer a more radical critique of the capitalist system as a whole[1]. This would result in more or less profound criticisms, and a split between one form of activism involved in the political sphere[2] and another, non-partisan one (Vrignon 2017).

Considering an electoral calendar which constrains and influences partisan discourse, we will focus here on non-partisan organizational forms, considered more independent. In a comparative approach, we assume that the diversity of contemporary ecologist organizations will reveal the construction of differing political identities and make it possible to distinguish language elements specific to forms of "ecologist subcultures", specific to various ecologist organizations.

The involvement of activists with info-communicational skills in these organizations would facilitate access to media arenas (Cefaï 2016), making it possible to form critical communities (Rabatel 2010), or even to acquire a form of media autonomy (Ferron et al. 2015). For example, it has been demonstrated that the Internet may constitute a space for activists to support forms of mobilization (Granjon 2014), notably through social media during strong mobilizations (Tufekci 2019).

The various actors in the ecologist struggle would be able to use different means of digital communication, particularly to try to defeat the framing effects of the dominant media (Sebbah 2019). This could notably be the case on social media, taking into account their role as infomediaries (Rebillard and Smyrnaios 2010, 2019)

1 Carbou, G. (2019). En finir avec la confusion entre écologie et environnementalisme. *Libération* [Online]. Available at: https://www.liberation.fr/debats/2019/06/16/en-finir-avec-la-confusion-entre-ecologie-et-environnementalisme_1734170/.

2 Embodied in 1974 by the presidential candidacy of René Dumont.

which could force ecologist organizations to maintain a presence there out of communicational necessity.

However, we consider that publicity on social media is subject to formal and substantive restrictions. The various maneuvers aimed at moderating the content of posts (Papaevangelou and Smyrnaios 2020) may, for example, lead to forms of censorship or self-censorship, like Facebook which blocked the accounts of Extinction Rebellion "rebels" in February 2020[3] or the spontaneous loss of audience by certain pages of the "radical left" between July and September 2019 (Badouard 2020)[4].

It is worth remembering that information and communications technologies (ICTs) could also do a disservice to the activists using them, whether due to the volatility of the mobilizations thus initiated (Tufekci 2019) or the potential infiltration of networks by agents of repression (Neveu 2019).

To try to establish ecologist themes as public problems, ecologists would therefore face a range of info-communicational channels supported by digital technologies, with advantages and disadvantages, which would contribute to making the register of digital actions more varied and complex (Van Aelst and Van Laer 2010). Ecologist organizations might thus use a variety of channels to compose specific media registers, like real media activists (Cardon and Granjon 2013). Organizations and their activists would thus be confronted with a "double rhetorical and publicization process" (Carlino and Stein 2019) in choosing both how to communicate and what to say. Among these channels, websites may be conducive to forms of liberated expression. The latter might seem essential to "exist on the Web" (Rouquette 2017) and would allow activists to publicize "content related to an editorial project" (Badouard 2014). It would be a more malleable channel, where organizations could deploy a freer discourse, less restricted than on social media. The website could then make it possible to find textual traces constituting self-published ecologist discourses.

---

3 Debove, L. (February 10, 2020). Alerte censure: tous les administrateurs de la page Extinction rebellion France ont été bannis de Facebook sans explications. *La Relève et la peste* [Online]. Available at: https://lareleveetlapeste.fr/alerte-censure-tous-les-administrateurs-de-la-page-extinction-rebellion-france-ont-ete-bannis-de-facebook-sans-explications/.

4 The apparent cause of which would be the sharing of so-called sensitive information, namely a photo of a police officer who had infiltrated activist networks.

13.2. Selection of organizations representative of the ecologist struggle in France and distinguishing their info-communicational methods

The constitution of our corpus begins from an exploratory phase both online, on the various info-communicational channels of the Web (websites and social media), and offline, during different ecologist mobilizations (section 13.5). From places of struggle to different social media, we identified 34 organizations, among which we distinguished three main types of organizations based on organizational criteria:

– traditional "associative" organizations, governed by the law of 1901;

– organizations that we will term "mixed and multipurpose" because they exploit several statuses;

– "autonomous and independent" organizations, without official status.

However, the boundaries between these categories remain permeable; they simply serve to distinguish between info-communicational practices according to the types of organizations. Then, we distinguished them according to their statuses, their fields of action, their levels of expertise and their means of communication in order to identify a new selection of 11 organizations in order to distribute them fairly across our three categories (see Table 13.1).

Categories	Name
Autonomous/independent	Amassada
	Bureburebure
	Radiaction
	Zad Nadir
Associative	Alternatiba
	Les Colibris
	France nature environnment
	Greenpeace
Mixed/multipurpose	ANV-COP 21
	Extinction Rebellion
	Notre Affaire à tous

Table 13.1. *Classification of selected organizations*

To represent the "autonomous and independent" organizations, we have therefore selected the organizations and sites of:

– Radiaction: the French version of the German collective *Ende Gelände*, defining itself as a "radical and disobedient ecologist collective against nuclear power and its world"[5];

– Amassada: a former Zone to Defend[6] (ZAD) in Aveyron, which stood in opposition to the installation of a very high voltage transformer and a wind farm;

– ZAD Nadir: corresponding to the former ZAD of Notre-Dame-des-Landes (NDDL);

– Bureburebure: also a former ZAD, occupied against the geological landfill project for nuclear waste in the Meuse.

Concerning associative organizations, we have chosen the sites of the following organizations:

– Alternatiba: an association of Basque origin now national in scope, offering alternatives to the dominant neoliberal system;

– Les Colibris: also a nationwide network of alternatives;

– Greenpeace: historic actor in the environmental struggle, particularly at the international level, that might also have been considered as a mixed and multipurpose actor through its proactive and multi-mediated communication;

– France Nature Environnement (FNE): historic federation bringing together numerous environmental associations.

As for "mixed and multipurpose" organizations, we selected:

– Extinction Rebellion (XR), a collective created more recently and omnipresent in a variety of locations of struggle;

– Action non violente-COP 21 (ANV-COP21): emerging from Alternatiba to implement actions of civil disobedience;

– Notre Affaire à tous: an association which mainly conducts legal actions.

They are governed by different statutes, act in distinct fields of action, both at local and national (French) level, show different levels of expertise and all have a

5 Introductory sentence on the site's home page: https://www.radiaction.org/.

6 Protest by occupation of areas dedicated to industrial projects.

website. They offer various strategies and means of communication (see Table 13.2) and thus comprise specific media repertoires.

Comparing the media repertoires and the info-communicational channels used by these organizations, we note that the "autonomous and independent" organizations use digital channels less than the "associative" and "mixed and multipurpose" organizations.

Subordinate collectives sometimes exist, such as organizations independent of the 11 selected, but their activity on social media is very weak or non-existent, apart from the "Bure à cuire!" collective. At the same time, all the "associative" and "mixed and multipurpose" organizations use Facebook, Twitter, Instagram and YouTube systematically.

		Info-communicational channels used								
Organizations		Site	Facebook	Twitter	Instagram	YouTube	Telegram	Mastodon	LinkedIn	Tik Tok
Auto./indep.	Amassada	x	x[7]							
	Bureburebure	x	x[8]	x[8]	x[8]		x[8]	x[8]		
	Radiaction	x	x	x			x[8]			
	Zad Nadir	x	x[8]							
Associative	Alternatiba	x	x	x	x	x	x	x[8]	x	
	Colibris	x	x	x	x	x		x	x	
	FNE	x	x	x	x	x			x	
	Greenpeace	x	x	x	x	x	x[8]	x	x	x
Mix. /multi	Anv-Cop21	x	x	x	x	x	x			x
	NAAT	x	x	x	x	x				
	XR	x	x	x	x	x	x	x	x	

Table 13.2. *Comparison of digital channels used by organizations*

7 An online presence exists, but managed by a single person or a collective subordinate to that which is the subject of the website in question; a former occupant of the ZAD for the case of Amassada, the collective "Bure à cuire!" for the case of Bure, a subordinate collective of the ZAD of NDDL, the local group "Alternatiba Nevers" for Alternatiba and a group dedicated to deforestation for Greenpeace.

8 The online presence in question is too old to be really significant (dated at least a year ago).

13.3. Textometric analysis of discourse on websites

Websites constitute self-mediatization channels allowing organizations to deploy a less constrained ecologist discourse than on social media. They thus allow us to access discourse imbued with the editorial brands of the 11 organizations. We consider it interesting to download the textual content in order to compose a corpus from the Web from which we can conduct a textometric analysis. To do this, we downloaded the textual content of each site using the Gromoteur scraping software[9] (Gerdes 2014). The latter downloads all the texts of the web pages whose URL addresses are given to it. In fact, it contains superfluous matter (banners, legal notices, GDPR forms, etc.) that need to be "cleaned up". An initial cleaning step is therefore necessary and can be carried out via the IRaMuTeQ[10] software, in order to exclude any superfluous content from the analysis. Then, we carry out a second step of textometric analysis, again with IRaMuTeQ, based on the remaining lexical forms making up the corpus.

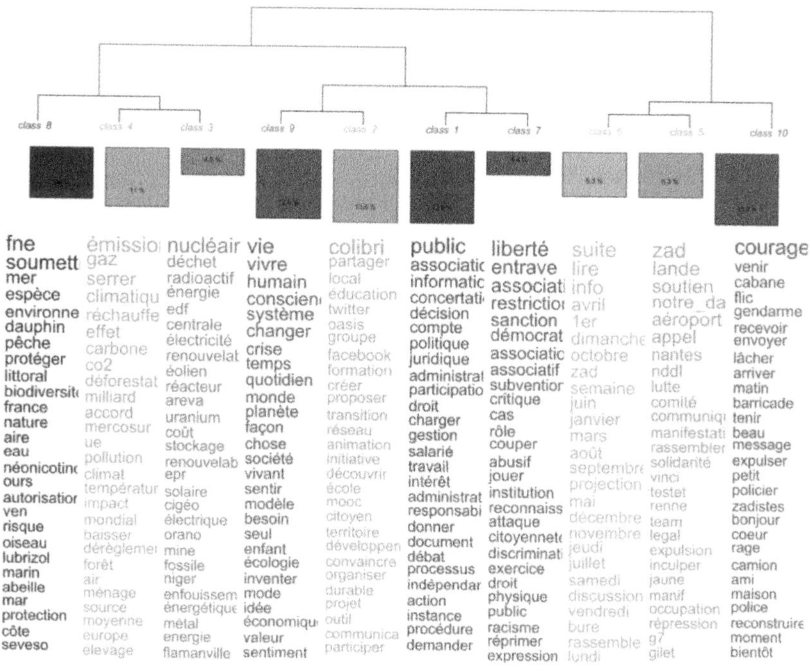

Figure 13.1. *Dendrogram of descending hierarchical classification For a color version of this figure, see www.iste.co.uk/catellani/transitions.zip*

9 This download took place in the period spanning July 2020 to December 2020.

10 This requires a second step of sorting, where recurring elements must be removed (top of page, tab names, thematic column, footer, etc.).

We thus obtain four families of themes for a total of 10 classes (see Figure 13.1):

– The first category comprises forms relating to the environment (class 8), fossil fuels (class 4) and nuclear energy (class 3).

– The second concerns the economic and ecological crisis (class 9), then the modalities of action of Les Colibris (class 2), the institutional system (class 1) and associative organizations (class 7).

– The third category includes forms related to current life in the ZAD of NDDL (class 6), the past struggles of the ZAD of NDDL (class 5), then the confrontations specific to ZADs (class 10).

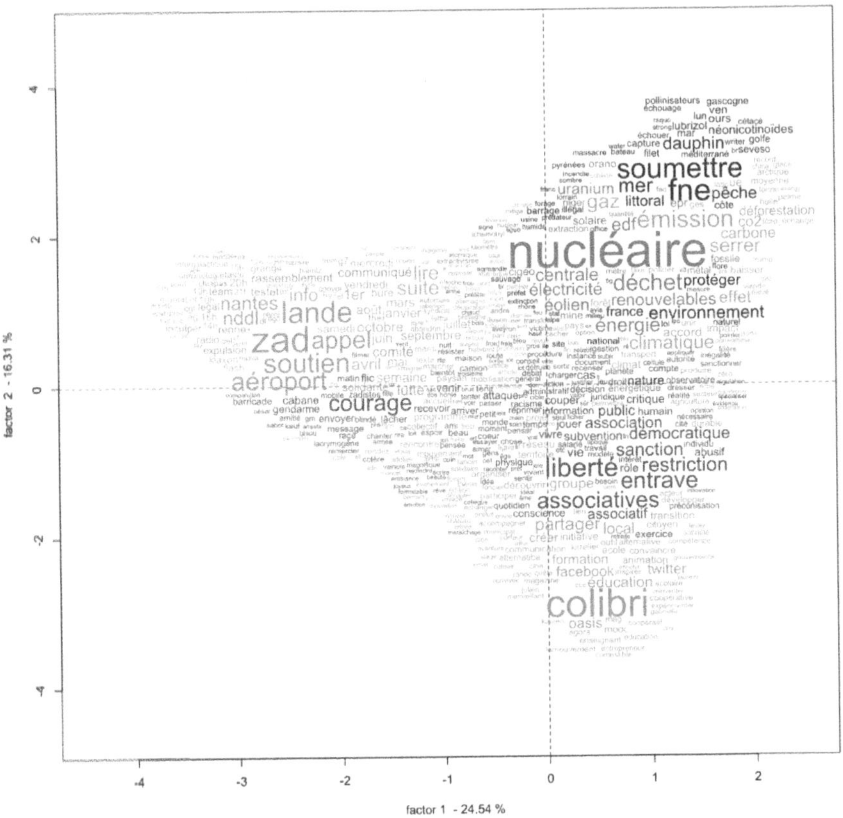

Figure 13.2. *Factorial analysis of correspondences between lexical forms For a color version of this figure, see www.iste.co.uk/catellani/transitions.zip*

A factorial analysis of correspondence (FAC, see Figure 13.2) allows us to determine the proximity between these themes and then to know which organizations relate to them. Thus, we notice that three types of discourse are distinguished by their singularity and give rise to a phenomenon of triangulation: the discourse of Les Colibris (in orange at the bottom right), that of ZAD Nadir (on the left), then those of FNE and Greenpeace (on the top right). The discourse of other organizations (Bureburebure, Radiaction, Amassada, Alternatiba, ANV-COP 21, Extinction Rebellion, Notre Affaire à tous) aggregate in the center, showing that they are semantically close.

The first observable phenomenon confirms the nebulosity of the ecologist movement and offers us clarification that strong divergences are active within it. "Les Colibris" advocate a moderate ecologism, through a sober way of life and an "ethical" capitalism. Next, ZAD Nadir offers two types of discourse, one more radical, which relates to past struggles in the ZAD (class 10) and the other more moderate, but still singular, advocating self-management as practiced daily on the ZAD (classes 5 and 6). Then, FNE offers a discourse approximating the environmentalist current, proposing a legalist ecologism, through the legislation of environmental norms (coastal law, anti-neonicotinoid law, regulation of fishing and hunting, etc.).

Finally, certain organizations share common criticisms, like the criticism of fossil fuels expressed by Alternatiba, ANV-COP 21, Extinction Rebellion, Greenpeace and Notre Affaire à tous (class 4). Nuclear energy is also criticized, but by fewer organizations (Bureburebure, Amassada, Greenpeace and Radiaction). In conclusion, the ecologist discourses studied are overall quite varied, ranging from environmentalism to more radical critique through discourses specific to certain modalities of action (ethical trade, self-management, etc.).

13.4. Discussion

"Associative" and "mixed and multipurpose" organizations are those which are the most active on the dominant social media. There, they seem to express a more environmentalist discourse, centered on certain aspects of the ecological struggle: anti-extractivism, the fight for the climate and the protection of biodiversity. Some also deploy themes specific to their operations, such as Les Colibris, who advocate their ethical consumption or FNE, which communicates about its environmental lobbying. Social media thus appear to be channels that are open to more moderate ecologist critiques.

The "autonomous and independent" organizations use social media[11] little and convey a rather anti-nuclear message, apart from ZAD Nadir, whose discourse is more singular[12]. Those most critical of nuclear energy are Bureburebure, a specialist in the fight against nuclear waste; Amassada, also focused on criticism of renewable energies; and Radiaction. They are supported by Greenpeace, whose criticism is nonetheless diluted in the variety of its discourse[13].

Finally, the info-communicational channels of the Web are indeed conducive to publicization of a diversity of ecologist discourses, but criticism of nuclear power does not seem to be very common on social media. Websites seem to be the info-communicational channels that host the greatest diversity of ecologist discourses, ranging from climate protection to anti-nuclear criticism.

However, we are presenting here only a non-exhaustive selection of ecological discourses, identified from online and offline observations. The ecologist critiques that emerge therefore depend largely on the organizations that we have selected in advance. These are only phenomena of mediatization of ecologist discourses, which indicate info-communicational trends. In fact, we conclude that the most radical criticisms are less common on social media, whereas websites constitute self-mediatization devices allowing more liberated speech.

In the future, it would be interesting to complement these analyses by integrating new organizations, both radical and moderate, to continue to observe the usefulness of this distinction between types of criticism and the info-communicational devices used.

Furthermore, the use of social media by the anti-nuclear movement would nevertheless seem to be making headway, as notably embodied by the presence of "Bure à cuire!" on social media. At a time when the latter seem gradually to occupy a hegemonic position in the information society, we wonder about the possibility of them hosting radical criticisms such as antinuclearism. For this, we suggest further investigation of the info-communicational channels of the Web to continue to elucidate which subjects can be debated in which media arenas.

---

11 Besides the subordinate collective of Bureburebure, "Bure à cuire!"

12 Specific to the past and present functioning of the former ZAD of NDDL.

13 Especially since in manually inspecting the corpus, we find textual traces attesting to management disagreements between these organizations, especially between Bureburebure and Greenpeace concerning the consultation mechanisms set up by the institutions.

13.5. Appendix: list of field observations

Locations	Organizations	Types of events
Toulouse	Alternatiba and ANV-COP 21	Information meetings, general assemblies, actions
	Alternatiba, ANV-COP 21 FNE	"Climate Marches"
Saint-Victor-et-Melvieu	Amassada	Rally and demonstration
Luméville-en-Ornois	Bureburebure	Festival
Paris	Alternatiba and ANV-COP 21	Work and reception places
Paris	Extinction Rebellion	Occupation

Table 13.3. *Locations and types of organizations observed during the exploratory phase*

13.6. References

Amossy, R. (2008). Argumentation et analyse du discours : perspective théorique et découpages disciplinaires. *Argumentation & analyse du discours. Revue électronique du groupe ADARR*, 1. doi: 10.4000/aad.171 [Accessed 12 January 2023].

Badouard, R. (2014). La mise en technologie des projets politiques. Une approche "orientée design" de la participation en ligne. *Participations*, 8(1), 31–54. doi: 10.3917/parti.008.0031 [Accessed 12 January 2023].

Badouard, R. (2020). *Les nouvelles lois du web. Modération et censure*. Le Seuil, Paris.

Bonnet, V. and Geslin, A. (2019). Les mots de l'écologie, 25 ans après. Circulation des discours et des notions. *Mots. Les langages du politique*, 119(1), 9–14. doi: 10.4000/mots.24186 [Accessed 12 January 2023].

Carbou, G. (2019). En finir avec la confusion entre écologie et environnementalisme. *Libération* [Online]. Available at: https://www.liberation.fr/debats/2019/06/16/en-finir-avec-la-confusion-entre-ecologie-et-environnementalisme_1734170/ [Accessed 12 January 2023].

Cardon, D. and Granjon, F. (2013). *Médiactivistes*. Presses de Sciences Po, Paris.

Carlino, V. and Stein, M. (2019). Militants et paroles militantes à l'épreuve des espaces de discussion. In *Les paroles militantes dans les controverses environnementales. Constructions, légitimations, limites*, Carlino, V., Stein, M. (eds.). Éditions de l'Universitaires de Lorraine, Nancy, 7–15.

Cefaï, D. (2016). Publics, problèmes publics, arènes publiques... Que nous apprend le pragmatisme ? *Questions de communication*, 2(2), 25–64. doi: 10.4000/questionsdecommunication.10704 [Accessed 12 January 2023].

Ferron, B. Harvey, N. and Trédan, O. (2015). *Des amateurs dans les médias. Légitimités, autonomie, attachements*. Presses des Mines, Paris.

Gerdes, K. (2014). Corpus collection and analysis for the linguistic layman: The Gromoteur. *Proceedings of the JADT*, Paris.

Granjon, F. (2014). Médias dominants, mouvements sociaux et mobilisations informationnelles. In *Histoire des mouvements sociaux en France de 1814 à nos jours*, Pigenet, M. and Tartakowsky, D. (eds). La Découverte, Paris, 681–687.

Lafon, B. (2019). *Médias et médiatisation. Analyser les médias imprimés, audiovisuels, numériques*. Presses Universitaires de Grenoble, Grenoble.

Neveu, É. (2019). *Sociologie des mouvements sociaux*. La Découverte, Paris.

Papaevangelou, C. and Smyrnaios, N. (2020) Régulation de la parole politique en ligne : les deux modèles antinomiques de Twitter et de Facebook. Ina [Online]. Available at: https://larevuedesmedias.ina.fr/regulation-discours-politique-twitter-facebook-reseaux-sociaux [Accessed 12 January 2023].

Rabatel, A. (2010). Analyse pragma-énonciative des s/citations du site d'Arrêt sur images. *Argumentation et analyse du discours. Revue électronique du groupe ADARR*, 4. doi: 10.4000/aad.763 [Accessed 12 January 2023].

Ratinaud, P. (2014). IRAMUTEQ, Interface de R pour les Analyses Multidimensionnelles de Textes et de Questionnaires. Un logiciel libre construit avec des logiciels libres [Online]. Available at: http://www.iramuteq.org [Accessed 12 January 2023].

Rebillard, F. and Smyrnaios, N. (2010). Les infomédiaires au cœur de la filière de l'information en ligne. Les cas de Google, Wikio et Paperblog. *Réseaux,* 160/161, 163–194. doi: 10.3917/res.160.0163.

Rebillard, F. and Smyrnaios, N. (2019). Quelle "plateformisation" de l'information ? Collusion socioéconomique et dilution éditoriale entre les entreprises médiatiques et les infomédiaires de l'Internet. *tic&société*, 13(1/2). doi: 10.4000/ticetsociete.4080 [Accessed 12 January 2023].

Reinert, M. (1983). Une méthode de classification descendante hiérarchique. Application à l'analyse lexicale par contexte. *Les Cahiers de l'analyse des données*, 8(2), 187–198.

Rouquette, S. (2017). Introduction. La stratégie d'un site internet, un exercice nécessaire. In *Site internet : audit et stratégie*, Rouquette, S. (ed). De Boeck Supérieur, Louvain-la-Neuve, 5–20.

Sebbah, B. (2019). *L'évènement politique en ligne*. Presses universitaires du Midi, Toulouse.

Tufekci, Z. (2019). *Twitter et les gaz lacrymogènes. Forces et fragilités de la contestation*. C&F Éditions, Caen.

Van Aelst, P. and Van Laer, J. (2010). Internet and social movement repertoires. *Internet Communication and Society*, 13(8), 1146–1171. doi: 10.1080/13691181003628307 [Accessed 12 January 2023].

Vrignon, A. (2017). *La naissance de l'écologie politique en France. Une nébuleuse au cœur des années 68*. Presses Universitaires de Rennes, Rennes.

14

Contrastive Analysis of Enunciation in Digital Citizen Participation

14.1. Introduction

From October 2019 to June 2020, the Citizens' Convention for the Climate[1] (hereinafter CCC) brought together 150 citizens drawn at random from the whole French population, to discuss measures to reduce greenhouse gas emissions with the goal of social justice. Mandated by the government, this participatory and deliberative mechanism benefited from strong media exposure before, during and after its occurrence[2]. At the same time, a website was launched allowing any person or organization to submit proposals to the Convention. These proposals, classified by the contributor into one of six suggested categories, were synthesized and forwarded to CCC members.

Without being directly connected to the CCC, from the beginning of April to the beginning of May 2020, a digital consultation in reaction to the Covid-19 pandemic was organized by a group of members of parliament known as *le Jour d'Après*[3] (The Day After; hereinafter JDA). This consultation, which aimed to imagine what post-pandemic life might be like, also benefited from some media exposure: the first lockdown had just been decreed and several of the initiators of the group had recently left the LREM parliamentary group associated with President Emmanuel Macron. The JDA consultation took place entirely online. Among the proposed themes, one in particular allowed the collection of contributions focused on adaptation to global warming (*consumption and sobriety*).

Chapter written by Antoine GABORIAU and Catherine DOMINGUÈS.

1 See: https://www.conventioncitoyennepourleclimat.fr/.

2 14,988 mentions in the press between January 1, 2019 and July 1, 2022 on Europresse.

3 See: https://lejourdapres.parlement-ouvert.fr/.

The objective of this work is to analyze how these two mechanisms of political expression used for consultations led the discursive communities thus constructed to produce enunciations that contrast on two points. The first concerns the handling of discourse. This involves quantitatively analyzing how each of the contributors says what they believe to be true (according to Culioli (1980)) by taking sole responsibility for their speech (*I*) or by associating it with a group (*we*). The second aims to identify the notions, people or objects designated by the enunciator as linked to them or to the groups they create in their discourse and to which they may or may not belong. For this, we analyze two enunciative traces (personal pronouns for the handling of discourse as well as possessive determinants used by the producer) that show how the contribution relates to its enunciation.

This research is thus situated in the field of discourse analysis defined as analysis of the relationship between text (contributions) and context (characterized here by the discursive community created by the mechanism). It seeks to interpret the regularities of the language observed in the contributions and the positioning and purposes specific to this same discourse, and thus in turn sustains a political interpretation of the two corpora.

The body of work consists of enunciations produced in real-life situations. The set of contributions from the CCC and those from the JDA's *consumption and sobriety* category constitute two sub-corpora whose numerical characteristics are summarized in Table 14.1.

	CCC	**JDA**
Contributions	3,445	1,709
Tokens[4]	468,207	259,430

Table 14.1. *Numerical characteristics of the two corpora studied. To facilitate reading, in all tables, boxes with a vertical-striped background refer to the CCC and boxes with a horizontal-striped background refer to the JDA*

Contributors to the two mechanisms form two communicative discursive communities in the sense that each of these people recognizes, in fact, the communicative mechanisms and contracts (Charaudeau 1983) characterized by:

– *identity traits of producers and recipients*: the enunciators/producers in the contributions are citizens in dialogue with others (CCC) or with elected officials (JDA);

4 A token is a lexical unit: a word, number or punctuation mark.

– *the purpose of the exchange*: "to define the structuring measures to achieve, in a spirit of social justice, a reduction of greenhouse gas emissions by at least 40% by 2030 compared to 1990" for the CCC, and "to exchange thoughts, share initiatives that work and concrete proposals to build the world of the future" for the JDA;

– *the thematic object*: ecology in different aspects of citizens' lives (housing, food, transport, etc.);

– *the circumstances of the exchange*: a website for both consultations.

These two discursive communities differ particularly in the identity traits of the recipients, differently constructed. For the CCC, this is an undifferentiated group characterized by desired heterogeneities of gender, age, skills, personal and professional situations, availability and so on. For the JDA, it is a group of elected people with similar sensibilities, identified as dissidents from the presidential majority. On the other hand, the enunciator is anonymized in both communities (biographical information given to the platforms is only declarative).

14.2. Production of the positioning of the enunciator

In both corpora, the enunciator is the producer of the contribution. They are the referent of the pronoun *I* and the source of the expressed point of view. But they may sometimes renounce the *I*, and thus give themselves "the possibility of associating with others" (Guespin 1985, p. 45). On this basis, the study of *we* in political texts is not new. In philosophy for a start, several academic works have shown the numerous uses of the term in the political field; one issue of a journal was even entirely devoted to it (Critique 2017). In linguistics, this fundamentally political question raised by the uses of *we* has been led in the field of discourse analysis by L. Guespin (1985).

The enunciation of *we* has been the subject of more limited studies in political science, in the French language. M. Offerlé and J. Frétel, for example, partially addressed this in their study of letters sent by citizens to the Élysée presidential palace (Frétel and Offerlé 2021). This is also the case for C. Durand in his study of letters from inmates to the penitentiary institutions (Durand 2014). Closer to our subject, M. Brugidou and his colleagues examine the use of the personal pronoun *we/us* in two corpora resulting from digital citizen participation mechanisms: the *Grand Débat National* (GDN, Great National Debate) and the *Vrai Débat* (Real Debate) (Brugidou et al. 2020; Suignard et al. 2021).

The construction of discursive positions in our two corpora can be analyzed through the way in which contributors position themselves in relation to the enunciations they produce. In the context of the designation of the enunciator, we are more precisely interested in the alternation *I/we* as it provides an angle for analysis of the link that the enunciator wishes to establish, through his or her different linguistic roles, with the recipients.

Morphosyntactic patterns were constructed in the Unitex[5] software suite in order to identify the sentence structures in the enunciations, which use personal pronouns of the first-person singular (*I/me*) and plural (*we/us*). Table 14.2 shows, in quantitative terms, the differentiated use of these pronouns in enunciations. In the CCC corpus, the referent pronoun of the first-person singular is more frequent than that referring to groups designated by the plural *we/us* (ratio 1st pers. sing./1st pers. plural: 1.51). In the JDA, the order of proportions is reversed and the referent pronoun *we* is more frequent than *I* (0.57[6]). Thus, an enunciator in the CCC makes proportionally more reference to their own experience, of which they assume the singularity, whereas in the JDA, they renounce this singularity and proportionally more frequently situate their contribution within *we/us*. This pronoun designates one or more groups, most often not characterized in the discourse. One interpretation might be that they try to "play on the ambiguity of the pronoun to draw the interlocutor towards this partisan *we*" (Guespin 1985) by including the recipients of this contribution in the same group as the speaker.

In a parallel manner, the enunciator may associate themselves with a group implicitly defined by their social role, of which they are not a part and often with a polemical aim, as in: "We were instantly able to mobilize 100 billion euros" (JDA), whose enunciator, after examining the entire contribution, is probably not a banker; or "We must stop bringing from the end of the world (garlic from Argentina, kiwi fruit from New Zealand, etc.) fruits and vegetables that we are able to grow in the metropolitan territory" (CCC). In the latter example, the enunciator may include him- or herself in the first *we* by association with economic actors who import food products from far away because these latter find consumers (including the speaker), but not in the second, unless the speaker is a farmer (which does not appear to be the case in light of the contribution).

---

5 See: https://unitexgramlab.org/fr and (Paumier et al. 2021).

6 For comparison, the journalistic corpus lm 94 (all publications of the newspaper *Le Monde* for the year 1994) gives a ratio of 1.72, which is close to that of the CCC, while the corpus of contributions to the great national debate concerning the ecological transition shows a ratio very different from the other three: 5.68.

Lemmas	CCC	JDA	lm94	GDN
1st person singular pronouns	1,202	553	37,046	424,222
1st person plural pronouns	794	833	21,478	74,696
Ratio of 1st pers. sing/ 1st pers. plural pronouns	1.51	0.66	1.72	5.68

Table 14.2. *Comparison of occurrences of 1st person singular and plural pronouns in different corpora. In addition to the previous convention (vertical-striped boxes referring to the CCC and horizontal-striped boxes referring to the JDA), the boxes corresponding to the two reference corpora have a gridded background*

Guespin (1985, p. 46 *sqq*.) insists on the multiplicity of possible referents of *we*, which, in discourse, always encompasses the enunciator *I*, but allows the addition of a second (that the enunciator may designate or describe in his or her contribution), classically an interlocutor (here, the recipients in the plural in both corpora), or even a non-person thus offering "a possible opportunity to give a position of co-enunciator to non-humans" (Guespin 1985).

This observation allows us to pose a hypothesis: are the different constructions of *we* produced by the enunciator found in the uses of the possessive determiner *our*? This involves questioning the link affirmed by this person in using the possessive determiner, and verifying whether the semantic fields to which these constructions refer contrast between the two corpora.

14.3. Objects of the discourse linked to the enunciator

Possessive determiners	CCC	JDA
1st pers. singular: *my*	937	416
1st pers. plural: *our*	1,420	1,360
Ratio of possessives 1st pers. sing./ 1st pers. plural	0.66	0.31
Ratio of possessives 1st pers. sing./ number of tokens	2	1.6
Ratio of possessives 1st pers. plural/ number of tokens	3.03	5.2

Table 14.3. *Comparison of occurrences of 1st person singular and plural possessives*

Possessive determiners make it possible to grammatically mark links between the objects and actors of the discourse. Syntax wise, these objects are designated by noun phrases determined by a possessive and constructed around a head noun that may be modified (as in the example *our common good*).

Table 14.3 shows the numerical significance of this designation of objects of discourse in both corpora.

In the two corpora, the 1st person singular possessive determiners (*my*) are less frequent than those of the 1st person plural (*our*) and the preponderance of the group possessive over the singular is, proportionally, more than twice as large in the JDA than in the CCC. For the JDA, this distribution of possessives (*our* instead of *my*) is consistent with that of subject pronouns (*we* instead of *I*): the contributor situating themselves within a group refers more often to objects linked to this group than to objects linked to the contributor.

On the other hand, the use of singular possessives is more common in the CCC corpus (2) than in the JDA (1.6). Contributors to the CCC therefore refer slightly more often than those in the JDA to links between the notions, people or objects designated in the discourse and the enunciator when the latter considers them in relation to themselves only (use of singular possessives). It is the opposite when the enunciator is part of a group (use of plural possessives).

The second part of the analysis focuses on analysis of the semantic fields to which these objects of discourse, shared between the enunciators and the recipients, may relate. We propose a typology broken down into three classes of our own design based on the referents of *we* suggested by Guespin (1985).

Let us first note the existence of an *our* marking a link between two communities. *Our farmers* falls into this class (1), since the expression implies that the speaker is not part of the agricultural community, but that they wish to connect their community of reference (the French people or a subset) with what they designate as the specific community of farmers. These examples of *our* create a link between two human communities. But the possessive can also establish a link which could be called the expression of a patrimonial *our* (2), between a human community and a "non-human" community, as in *our forests*. These designations take on a hypocoristic value in that they aim to create or highlight a bond of familiarity or connivance between the co-enunciators who form the community of reference, through a shared view toward another community.

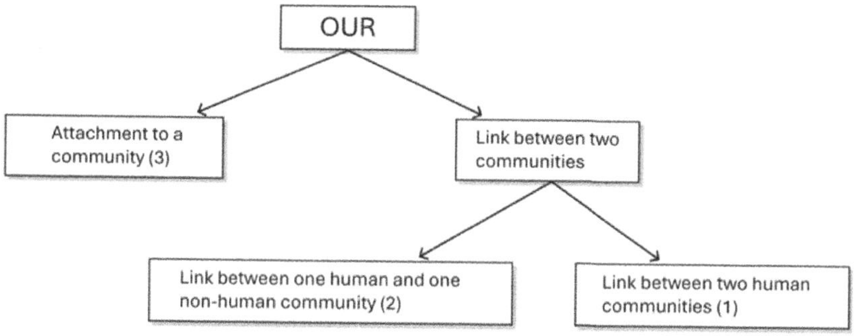

Figure 14.1. *Summary diagram of the proposed typology of analysis*

We construct the third class of analysis of *our* as that of an encompassing *our* (3), expressing the idea of a single community to which the contributing person expresses their attachment, with which they share the objects of discourse and therefore the points of interest and concerns. For example, we find in this class the expression *our needs*. These expressions do not refer to material questions (as opposed to the link between a human community and a non-human community), and do not establish a link between two human communities.

It should be noted that this typology does not suggest fixed boundaries between different classes; the same expression may, depending on the occurrences, be classified in one category or another. The example *our planet* is edifying: this expression may be linked to the patrimonial *we* (2) in certain contexts, and to the *we* of attachment to a community (3) in others. The most frequent noun expressions ("proportion" (see note 6) higher than 2.5) were classified according to this typology. Due to lack of space, Table 14.4 shows only those with a frequency greater than 3.

	CCC	Rank in CCC	JDA	Rank in JDA	Class in the typology
Terms classified at similar ranks in the two corpora					
Consumption	14.31	1	18.12	1	3
Planet	9.82	2	17.35	2	2
Lifestyle	5.13	8	8.09	8	3
Waste	4.49	10	7.32	10	2
Economy	3.42	12	7.32	11	3
Terms whose classifications differ from 3 to 6 ranks between the two corpora					

	CCC	Rank in CCC	JDA	Rank in JDA	Class in the typology
Environment	5.34	7	11.56	4	2
Country	8.33	3	10.02	6	2
Child	7.90	4	8.48	7	1
Terms whose classifications differ by 7 or more ranks between the two corpora					
Need	4.49	9	11.95	3	3
Mode	1.28	6	6.17	13	3
Emission	7.69	5	5.01	14	2
Terms classified for the JDA corpus but not passing the threshold in the CCC corpus					
Farmer	1.92		11.18	5	1
Food supply	1.71		7.71	9	3
World	2.35		6.17	12	2
Model	1.71		5.01	15	3
Producer	0.00		4.63	16	1
Habits	2.35		3.85	17	3
Farmer	1.92		3.47	18	2
Good	0.43		3.47	19	3
Capacity	1.71		3.08	20	3
Fellow citizen	2.14		3.08	21	1
Space	0.64		3.08	22	2
Housing	0.64		3.08	23	2

Table 14.4. *Classification, in the typology of analysis, of the head nouns of the most frequent noun phrases determined by the lemma "our" with their rank in the CCC corpus, then in the JDA corpus*

We then proceeded to count these classifications in order to compare the use of *our* in the two corpora. Table 14.5 shows these counts and includes the proportions.

	1 (*proportion*)	2 (*proportion*)	3 (*proportion*)
CCC	1 (*8%*)	6 (*46%*)	6 (*46%*)
JDA	6 (*20%*)	9 (*30%*)	15 (*50%*)

Table 14.5. *Distribution of notions, persons or objects designated in the discourse and shared by the enunciator in the proposed typology of analysis*

The CCC corpus is marked by less use of *our* as a connection between two human communities, while the use of *our* as a link between a human and a non-human community prevails. This suggests that more contributors to the CCC would like to draw attention to concrete problems (*waste*, *housing*, *environment*, *emissions*, etc.) for which responsibility should be collective. Contributors to the JDA focus more, in comparison, on establishing links between two human communities (our farmers, our politicians, our young people, our children, etc.), and therefore on drawing attention to specific human sub-communities, perhaps to be protected.

14.4. Conclusion

We have therefore described here the handling of enunciation in two corpora of political texts that are citizens' contributions to a digital consultation. Our study focused initially on the enunciating designations *I* and *we*, then shifted to the objects of speech linked, by means of a possessive determiner, to these enunciators. We then proposed a typology of analysis for these objects, enabling comparison of the two corpora studied. Thus, if the recipients of the contributions are other citizens, it seems that the contributor will be more inclined to speak on their own behalf and will share their personal experience. If the recipients are elected officials, on the other hand, the contributor will tend to refer in their discourse to experiences associated with one or more groups, which illustrates the importance of the prior framing of the mechanism – and more specifically the attention paid to the mobilized speakers – for situations of enunciation.

Replication of this type of study would firstly make it possible to confirm (or refute) the link that we create between enunciators and recipients, as well as providing an overall view of the methods of framing of digital participation mechanisms and the adaptation of enunciation which accordingly results, thus extracting itself from the pre-existing norms of communication within this type of mechanism. It would also allow better knowledge and understanding of this new type of text, which has appeared with digital consultation mechanisms.

14.5. References

Brugidou, M., Suignard, P., Escoffier, C., Charaudeau, L. (2020). Un discours et un public "Gilets Jaunes" au cœur du Grand Débat National. Combinaison des approches IA et textométriques pour l'analyse de discours des plateformes "Grand Débat National" et "Vrai débat". *JADT, 2020 : 15e Journées internationales d'analyse statistique des données textuelles*.

Charaudeau, P. (1983). *Langage et discours. Éléments de sémiolinguistique*. Hachette, Paris.

Charaudeau, P. and Maingueneau, D. (2002). *Dictionnaire d'analyse du discours*. Le Seuil, Paris.

Critique (2017). Nous. *Critique*, 144, 841–842 [Online]. Available at: https://www.cairn.info/revue-critique-2017-6.htm.

Culioli, A. (1980). Valeurs aspectuelles et opérations énonciatives : l'aoristique. *Recherches linguistiques*, 5, 182–193.

Durand, C. (2014). Construire sa légitimité à énoncer le droit. Étude de doléances de prisonniers. *Droit et société*, 87(2), 329–348. doi: 10.3917/drs.087.0329.

Frétel, J. and Offerlé, M. (2021). *Écrire au président : enquête sur le guichet de l'Élysée*. La Découverte, Paris.

Guespin, L. (1976). Typologie du discours politique. *Langages*, 41(10).

Guespin, L. (1985). Nous, la langue et l'interaction. *Mots. Les langages du politique*, 10(1), 45–62. doi: 10.3406/mots.1985.1184.

Leroy, S. (2004). *Le nom propre en français*. Ophrys, Paris.

Paumier, S., Flury, W., Guenthner, F., Laporte, E., Malchok, F., Marschner, C., Vollant, G. (2021). UNITEX 3.3 User Manual [Online]. Available at: https://hal.science/hal-03589580 [Accessed 13 January 2023].

Riegel, M., Pellat, J.C., Rioul, R. (1994). *Grammaire méthodique du français*. Presses Universitaires de France, Paris.

Suignard, P., Escoffier, C., Charaudeau, L., Brugidou, M. (2021). Que peuvent les algorithmes de plongement de mots pour l'analyse sociologique des textes ? Analyser les discours et caractériser les locuteurs des plateformes "Grand Débat National" et "Vrai Débat". *Statistique et société*, 9(1/2), 133–145 [Online]. Available at: http://statistique-et-societe.fr/article/view/825.

15

Political Mediations in the Case of Energy Transition: Deep Geothermal Energy as Perceived by Local Residents

Public policies of "energy transition" in Alsace have led to a multiplication and acceleration of deep geothermal energy (DGE) power plant projects. Since the 1980s, these technologies have been successfully tested at experimental sites and then set up in rural areas. However, since 2014, the development of new projects in the Strasbourg Eurometropolis (SEM), with a high population density, has contributed to the emergence of a socio-technological controversy. More than a "purely" technical dimension, this concerns the territorial location of projects and methods of decision-making which condition and restrict public participation in these policies. Many criticisms thus explicitly referred to a technocratic mode of decision-making (Habermas 1973) in a situation of risk.

DGE, sometimes referred to as a "non-mature technology" (Serrano et al. 2019), may involve risks, particularly in terms of induced seismicity. And while forms of protest could be described as "irrational", or even reduced to a hypothetical "NIMBY" syndrome (Chavot et al. 2019b) by the public authorities, these risks have nevertheless materialized in significant seismic events of an anthropogenic nature, caused by drilling on the Reichstett site, since November 2019 (Chavot et al. 2021). And projects authorized by the prefecture, often contrary to the opinions produced during public inquiries, will then be frozen.

On a theoretical level, public communication reduced to a policy of popularization has frequently contributed to depoliticization of controversies and

---

Chapter written by Cyrille BODIN, Philippe CHAVOT, Anne MASSERAN, Yeny SERRANO and Jean ZOUNGRANA.

public debates in situations of risk, by constraining the expression of hybrid expertise, taking technological knowledge into consideration through its social, cultural, environmental, economic or heritage particularities (Bodin et al. 2018; Bodin 2019). Also, a division arising from a positivist political tradition between "technology and territory, engineering logic and emotion, economy and love, reason, and passion […], general interest and particular interests [...] must be reclassified" (Fourniau 2007, p. 73) in order to understand the frequently expressed opposition towards territorial environmental development projects.

This study is part of a long-term analysis of the controversy, conducted using complementary approaches (analysis of public and media discourses, analysis of participation or consultation mechanisms, analysis of public perceptions and social mobilizations, analysis of stakeholder strategies). A plurality of "participation" mechanisms (Carlino and Stein 2019), devoid of any decision-making content, have notably been deployed at different political-administrative levels (public inquiries, site monitoring committees, information and evaluation missions), without however completely responding to the arguments put forward by some of the populations, associations and certain local elected officials.

To complement the prior work of our team, previously cited, six focus group sessions were thus organized between January and February 2018, in order to allow a qualitative-type analysis of the positions of populations residing near power plants or projects for future power plants. The sessions were attended by 28 participants, invited on both a voluntary and random basis (invitation from a quantitative survey administered face to face in the street, quota method, n = 881). During the sessions, we successively addressed four themes offering scenarios relating to modes of energy production and consumption, to their ecological transition as well as to an effective, largely hypothetical, form of public participation in the territories.

The use of focus groups conducted among residents presents several methodological particularities which seem to us to be suited to the study of representations linked to DGE, particularly in the context of a controversy where resident voices remain little heard and even less considered in the implementation of public policies. First of all, the semi-directed and pragmatic character of this method circumvents a radical opposition created between the figure of the "citizen" supposed to represent a general interest, and that of the "resident" deemed to be locked into a particular interest (Fourniau 2007). By letting participants express, through an "unfolding" discussion, their own vision of the world and their own identity, focus groups offer freedom of speech allowing a move beyond the usual games of labeling, often mobilized for the purposes of political disqualification.

> Issues articulating identity and more or less standardized collective representations are particularly well served by the developmental

> dimension of the focus group. The group will then itself allow the emergence of different representations and interpretations according to each person's identity. The discussion will allow us to see whether consensus is possible or not (Masseran and Chavot 2006, p. 145).

Subsequently, this data collection technique allowed us to generate debates and exchanges of arguments according to some of the criteria of political participation: this method is based on qualitative representativeness of the participants, based on the criteria of voluntary, free and supportive commitment (Dacheux 2005). It should also be emphasized that this research procedure allows the creation of knowledge for both researchers and interviewees, a characteristic that we will discover frequently during the sessions, when, notably, the participants adopt a reflexive position regarding their own respective knowledge.

Finally, this method allows participants to intervene in cases, here semi-fictitious, allowing construction of a collective projection. Less restrictive than a discussion about already-existing experiences in individual life-worlds, focus groups go beyond the simple framework of the "life story" to, on the contrary, promote social projection through an active and collective exchange of arguments. They also help identify possible changes in individuals' positions during the discussion. In this context, this method remains resolutely exploratory and inductive: it allows researchers to develop new hypotheses, to cross-reference data obtained from other research methods and to evaluate the public reception of institutional discourses (Touré 2010).

Taking place at the crossroads of the analysis of environmental controversies and socio-scientific mediations, this work allows us to consider anew territorial political mediations concerning the energy transition and their relationships to forms of knowledge. Also, while public controversy appears both inevitable and certainly indispensable, we sometimes observe that participants in focus groups desire an inclusive diversification of research groups, implying at the same time recourse to public debate, as well as to participatory and citizen research (Catellani et al. 2021). We will therefore, in conclusion, reflect on territorial political mediations and their deleterious effects in terms of the energy or more generally the environmental transition.

15.1. Evaluating the benefits and risks of deep geothermal energy

Our first question concerned the profiles of the participants engaged in the procedure. The 28 participants present socio-demographic characteristics with relatively homogeneous tendencies: people of rather advanced age (on average 62 years), predominantly male (only eight women) and working in or having mainly worked in high-tech professions (engineer, specialized technician, public sector scientist, R&D or banking executive, business manager, etc.), most often requiring higher education diplomas.

The risks discussed during the sessions mainly concern the first three major stages in the life of DGE plants: drilling, stimulation, then operation of the plant. Questions relating to dismantling production units at the end of their life were addressed more rarely. Although multiple, the two main risks mentioned are related to induced seismicity (or modification of the topography of the soil) and to reduced quality (or availability) of water resources. In certain discussions, these categories of risks are sometimes linked: excessive and uncompensated extraction from the water table could, according to some participants, lead to land subsidence.

On the other side of the balance, the benefits mentioned during the discussions are just as numerous. These sometimes incorporate special or local interests, for example, by providing residents of the municipalities concerned with lower-cost heating. But they also appear to be conceived according to a common interest. From an environmental point of view, it obviously entails participating in the efforts required to carry out an energy transition and to reduce CO_2 emissions.

Two main reasons were given by participants. Firstly, DGE is a non-intermittent energy source that, unlike photovoltaics or wind power, avoids climatic and weather hazards. Secondly, the benefits mentioned also relate to the local economy: it will increase the value of resources conceived within the framework of territorial competitiveness, and make it possible to generate local jobs. And participants in the focus groups, after an exchange of arguments, always make a positive collective evaluation of the balance of benefits and risks linked to DGE developments in general: “It’s a beautiful energy!” one participant told us. “You get nothing for nothing, you have to give it a try!” said another.

However, what generally appears to be a favorable collective assessment of DGE and its uses is usually accompanied by conditions related to territorial management and development. Frequently, participants mention the need to build production units a good distance away from urban areas, and sometimes also from rivers or the water table. Generally speaking, oppositions relate less to the technology as such than to developments of DGE in its particular contexts, notably in its choices of territory or in the political modalities of its local deployment. It is then a question of diversifying the composition of the collectives responsible for implementing projects, by authorizing a socio-technological negotiation between technical expertise seen to be too “generic” and territorial expertise allowing more specific considerations to be introduced.

15.2. The challenges and shortcomings of a policy of popularization

The issue of publicizing information on DGE often emerges spontaneously, even before we address the fourth theme relating to the implementation of citizen participation in a territory's energy policy. The issues then raised by the discussion are most frequently those of availability of information, its reliability, as well as its independence, in order to allow collective decision-making and responding to a public interest. The question of information, before any political decision-making, is obviously presented as the initial precondition for engagement in local public life.

The sources of information used by the participants during the sessions appear to be particularly rich and diverse. These remain variable in their forms, however, depending on the individuals and their positions. A participant will come to one of the meetings with a printed list, "from the Internet", of the chemicals used by manufacturers in hydraulic fracturing, in order to support an argument concerning the harmful environmental consequences of these technologies. Among the sources used, we also note the mention by a participant, at the presentations, of internal discussions specific to the residents' association of the Poteries district in Strasbourg, involved in the development of a project in Eckbolsheim, at the Strasbourg border (Bodin et al. 2018). We also note the use by participants with technical or scientific professional profiles of literature from specialized institutions (BRGM, ADEME, etc.). In this context, documentary resources from public research institutions seem to enjoy a credibility that, conversely, resources from the general media or local political institutions do not. These observations appear to largely corroborate the results of our earlier quantitative research (Chavot et al. 2019a).

Also, the argumentative dynamics seen in the discussions never show a firm and definitive rejection of DGE. On the other hand, they imply rather the use of great caution: the technology is generally considered "promising", but nevertheless remains fraught with risks deserving special attention and guarantees. Thus, some of the self-declared most cautious participants will gradually clarify their positions during the discussion, bringing forward multiple arguments, sometimes on the limits of State guarantees against mining disasters, sometimes on local risks and ways of preventing them, sometimes on potential misuses of the technology by operators. The social relationships constituting the development of the technology itself are then under particular scrutiny: complex relationships between the State and the territories, relationships of dependence and exacerbated economic competition between local communities, relationships of distrust displayed by residents against elected officials and the local general news media, also relationships of economic dependence between public research institutions and industrial operators.

And finally, while the political, institutional and media world is pointed towards its own organizational or discursive contradictions and inconsistencies, some participants propose the constitution of "citizen commissions" to generate local (counter-)expertise, to be directly mobilized during socio-technological negotiations, according to inclusive and long-term logics, in order to adapt the technologies to the territories.

15.3. Conclusion: reinventing territorial political mediations of the energy transition?

Thus, from the point of view of the interviews carried out and our analyses, opposition to the development of DGE in Alsace seems to relate less to "purely" technical questions, than to the difficulty experienced by citizens in being able to publicly discuss the merits of projects, with a view to producing a collective decision appropriate to the territories. In this context, local considerations of politics, economics, culture or heritage, as well as sometimes geology, are all likely to come into play on a case-by-case basis depending on the projects and territories.

Although the discussions produced show a certain "localist" idealism of participation and democracy (Gadras and Pailliart 2013), criticism primarily relates to a technocratic and centralized mode of decision-making, which contributes to the generation of controversy in its most conflicting aspects. Public information in this area seems essential and at the same time insufficient. When the project presents a situation of risk for the territory and the community, an absence of deliberation following the public being informed can clearly lead to a kind of "radicalization" of the challenge. Conversely, considering that controversy is indispensable both to democratic life and to the construction of scientific or technical knowledge, it is perhaps better to promote it so as to facilitate the exchange of arguments in a pluralistic, symmetrical and open way (Pestre 2007).

This change of perspective in the conduct of local political mediations, in terms of territorial development and implementation of the energy transition, leads to two main reflections. Firstly, public mediations conducted according to the sole perspective of an information campaign, quickly reduced to a policy of popularization, seem largely counterproductive (Bodin 2019). If they help to temporarily depoliticize the controversy by hindering debate on projects, they contribute at the same time, in the long term, to strengthening and hardening social mobilization. And at the same time, they permanently discredit technologies that are generally considered beneficial to the environment and society. Secondly, in the opinion of some participants, a change in the conduct erritorial political mediations leaning toward effective participation, both in terms of public debate and citizen research, would also support citizen engagement and collective responsibility at a local level. However, both could greatly enhance the effectiveness of public action in the field of environmental policy.

A public debate and recourse to territorialized participatory research could thus allow a meeting "between, on the one hand, the familiar world of experience and, on the other hand, autonomous and specialized institutional subsystems, which find it difficult to manage this uncertainty and the crises resulting from it" (Pascual Espuny 2014). Therefore, if an energy or environmental transition policy is to be effective and socially responsible in the long term, it should also be based on a properly communicational and epistemological transition (Habermas 1973; Beck 1986), leading the public authorities to reject the simplicities and inconsistencies of a relationship of political domination supposedly conducted "in the name of science".

15.4. References

Beck, U. (1986). *La société du risque. Sur la voie d'une autre modernité*. Flammarion, Paris.

Bodin, C. (2019). La vulgarisation contre la participation, les paroles "militantes" au regard de la CNDP Nanotechnologies. In *Les paroles militantes dans les controverses environnementales. Constructions, légitimations, limites*, Carlino, V., Stein, M. (eds). Éditions de l'Universitaires de Lorraine, Nancy.

Bodin, C., Chavot, P., Masseran, A., Serrano, Y., Heimlich, C., Zoungrana, J. (2018). Les formes d'appropriation de la transition énergétique par le récit territorial : le cas de la géothermie profonde dans la métropole de Strasbourg. *Les enjeux de l'information et de la communication*, 18/3A, 69–80. doi: 10.3917/enic.hs6.0069.

Carlino, V. and Stein, M. (eds) (2019). *Les paroles militantes dans les controverses environnementales. Constructions, légitimations, limites*. Éditions de l'Universitaires de Lorraine, Nancy.

Catellani, A., Pascual Espuny, C., Jalenques-Vigouroux, B. (eds) (2021). Les sciences participatives au prisme des sciences de l'information et de la communication. *Études de communication*, 56. doi: 10.4000/edc.11315.

Chavot, P., Bodin, C., Heimlich, C., Masseran, A., Serrano, Y., Zoungrana, J. (2019a). Public perception of geothermal projects in Alsace: Between energy transition and territorial preservation. *European Geothermal Congress 2019*, 11-14 June [Online]. Available at: https://europeangeothermalcongress.eu/wp-content/uploads/2019/07/284.pdf.

Chavot, P., Masseran, A., Serrano, Y., Bodin, C., Heimlich, C., Zoungrana, J. (2019b). "Nimby et fier de l'être" : identité, étiquette et visions du territoire. In *Les paroles militantes dans les controverses environnementales. Constructions, légitimations, limites*, Carlino, V. and Stein, M. (eds). Éditions de l'Universitaires de Lorraine, Nancy.

Chavot, P., Masseran, A., Serrano, Y., Zoungrana, J. (2021). La science est-elle soluble dans des projets techno-industriels ? Querelles autour de la géothermie profonde à l'Eurométropole de Strasbourg. *Les enjeux de l'information et de la communication*, 22(4), 63–76.

Dacheux, E. (2005). *Associations et communication. Critique du marketing*. CNRS Éditions, Paris.

Fourniau, J.-M. (2007). "Citoyen en tant que riverain" : une subjectivation politique dans le processus de mise en discussion publique des projets d'aménagement. In *Le débat public : une expérience française de démocratie participative*, Blatrix, C. et al. (eds). La Découverte, Paris.

Gadras, S. and Pailliart, I. (2013). Les territoires et les médias dans la construction de l'espace public. In *Médias et territoires. L'espace public entre communication et imaginaire territorial*, Noyer, J., Raoul, B., Pailliart, I. (eds). Presses Universitaires du Septentrion, Villeneuve-d'Ascq.

Habermas, J. (1973). *La technique et la science comme "idéologie"*. Gallimard, Paris.

Masseran, A. and Chavot, P. (2006). Un artefact participatif : le focus group, potentialités et limites. In *La communication organisationnelle en question : méthodes et méthodologies*, Bouzon, A. and Meyer, V. (eds). L'Harmattan, Paris.

Pascual Espuny, C. (2014). La société civile, de l'alerte à la controverse médiatisée. *Communication et organisation*, 45, 115–126. doi: 10.4000/communicationorganisation.4531.

Pestre, D. (2007). L'analyse de controverses dans l'étude des sciences depuis trente ans. Entre outil méthodologique, garantie de neutralité axiologique et politique. Mil neuf cent. *Revue d'histoire intellectuelle*, 25, 29–43. doi: 10.3917/mnc.025.0029.

Serrano, Y., Heimlich, C., Bodin, C., Chavot, P., Masseran, A., Zoungrana, J. (2019). La géothermie profonde "n'est pas mature" : parcours d'une formule-argument à l'Eurométropole de Strasbourg. *Mots, les langages du politique*, 119, 51–67. doi: 10.4000/mots.24318.

Touré, E.H. (2010). Réflexion épistémologique sur l'usage des focus groups : fondements scientifiques et problèmes de scientificité. *Recherches qualitatives*, 29(1), 5–27. doi: 10.7202/1085130ar.

List of Authors

François ALLARD-HUVER
CREM
University of Lorraine
Nancy
France

Pauline AMIEL
IMSIC
Aix-Marseille University
University of Toulon
France

Cyrille BODIN
GRESEC
Grenoble Alpes University
France

Andrea CATELLANI
RECOM, LASCO
UCLouvain
Belgium

Mikaël CHAMBRU
GRESEC
Grenoble Alpes University
France

Philippe CHAVOT
LISEC
University of Strasbourg
France

Cécilia CLAEYS
CRESEM
University of Perpignan Via Domitia
Perpignan
France

Amélie COULBAUT-LAZZARINI
GRESEC
Grenoble Alpes University
France

Quentin DAVEAU
GRESEC
Grenoble Alpes University
France

Orélie DESFRICHES-DORIA
Paragraphe
Paris 8 University
Saint-Denis
France

Jules DILÉ-TOUSTOU
LERASS
Toulouse III –
Paul Sabatier University
Castres
France

Jean-Claude DOMENGET
ELLIADD
University of Franche-Comté
Montbéliard
France

Catherine DOMINGUÈS
LASTIG
Gustave Eiffel University
Saint-Mandé
France

Antoine GABORIAU
CESPRA
EHESS
Paris
France

Bruno LEFÈVRE
LabEx ICCA
LabSIC
Sorbonne Paris Nord University
France

Grégoire LITS
ORM
ILC
UCLouvain
Belgium

Catherine LONEUX
PREFics
Rennes 2 University
France

Anne MASSERAN
CREM
University of Strasbourg
France

Jeanne PAHUN
LISIS
INRAE
Champs-sur-Marne
France

Céline PASCUAL ESPUNY
IMSIC
Aix-Marseille University
France

Catherine QUIROGA CORTÉS
LERASS
Toulouse III –
Paul Sabatier University
France

Erika RIBERI
IMSIC
Aix-Marseille University
University of Toulon
France

Brigitte SEBBAH
LERASS
Toulouse III –
Paul Sabatier University
France

Yeny SERRANO
LISEC
University of Strasbourg
France

Natacha SOUILLARD
IMSIC
Aix-Marseille University
France

Bi Mathieu TRA
ELLIADD
University of Franche-Comté
Montbéliard
France

Kimberley VANDENHOLE
SONYA
Free University of Brussels
Brussels
and
Research Center for Sustainable Development
Ghent University
Belgium

Albin WAGENER
ICAM
PREFics
Rennes 2 University
France

Jean ZOUNGRANA
SAGE
University of Strasbourg
France

Index

L, M

N, P

R, S

T, U, W

Index

L, M

N, P

R, S

T, U, W

Other titles from

iSTE

in

Science, Society and New Technologies

2024

BARTHES Angela, CIBIEN Catherine, ROMAGNY Bruno
Biosphere Reserves and Sustainable Development Goals 1: Scientific and Practical Educational Issues in the Mediterranean
(Territory Development Set – Volume 3)
Biosphere Reserves and Sustainable Development Goals 2: Issues, Tensions, Processes and Governance in the Mediterranean
(Territory Development Set – Volume 4)

DAVALLON Jean
Heritage Traces in the Making: A Communicational Analysis of Modes of Heritagization
(Traces Set – Volume 6)

DEJEAN Maxime, L'HOSTIS Marine, LI Meng
Chinese Outbound Tourism: From Source Markets to Destinations
(Tourism and Mobility Systems Set – Volume 8)

GUILHEM Julia, FAUCHON Anne, KANAWATI Rushed
Al, Healthcare and Law

KOHLMANN Émilie
Linking with Nature in the Digital Age
(Communication, Environment, Science and Society Set – Volume 1)

LEDOUBLE Hélène
Popularizing Science: The Complex Terminological Interactions between Scientific and Press Discourses within the Field of Agroecology
(Communication, Environment, Science and Society Set – Volume 2)

LOUART Pierre, PIRIOU Jérôme, VIOLIER Philippe
Tourist Destinations According to Stakeholder Strategies: From Clusters to Local Tourist Systems
(Tourism and Mobility Systems Set – Volume 7)

2023

APCHAIN Thomas
Favela Tours: Building Otherness in the Favelas of Rio de Janeiro
(Tourism and Mobility Systems Set – Volume 6)

CARDON Alain, EL HAMI Abdelkhalak
Fundamental Generation Systems: Computer Science and Artificial Consciousness, the Informational Field of Generation of the Universe, the Sixth Sense of Living Beings
(Digital Sciences Set – Volume 4)

CURULLI Irene G., IKIZ KAYA Deniz, KHAEFI Arghavan
Heritage Education for Climate Action
(Research in Architectural Education Set – Volume 2)

ELAMÉ Esoh
The Sustainable City in Africa Facing the Challenge of Liquid Sanitation
(Territory Development Set – Volume 2)

JURCZENKO Emmanuel
Climate Investing: New Strategies and Implementation Challenges

LYUBAREVA Inna, WALDECK Roger
Diversity of Methodological Approaches in Social Sciences: Example of the Analysis of Media and Online Information
(Modeling Methodologies in Social Sciences Set – Volume 3)

PARET Dominique, CREGO Pierre, SOLERE Pauline
Smart Patches: Biosensors, Graphene and Intra-Body Communications

VENTRE Daniel, GUILLOT Philippe
Electronic Communication Interception Technologies and Issues of Power

VENTRE Daniel, LOISEAU Hugo
Cybercrime during the SARS-CoV-2 Pandemic (2019-2022): Evolutions, Adaptations, Consequences
(Cybersecurity Set – Volume 3)

2022

AIT HADDOU Hassan, TOUBANOS Dimitri, VILLIEN Philippe
Ecological Transition in Education and Research

CARDON Alain
Information Organization of The Universe and Living Things: Generation of Space, Quantum and Molecular Elements, Coactive Generation of Living Organisms and Multiagent Model
(Digital Sciences Set – Volume 3)

CAULI Marie, FAVIER Laurence, JEANNAS Jean-Yves
Digital Dictionary

DAVERNE-BAILLY Carole, WITTORSKI Richard
Research Methodology in Education and Training: Postures, Practices and Forms
(Education Set – Volume 12)

ELAMÉ Esoh
Sustainable Intercultural Urbanism at the Service of the African City of Tomorrow
(Territory Development Set – Volume 1)

FLEURET Sébastien
A Back and Forth Between Tourism and Health: From Medical Tourism to Global Health
(Tourism and Mobility Systems Set – Volume 5)

KAMPELIS Nikos, KOLOKOTSA Denia
Smart Zero-energy Buildings and Communities for Smart Grids
(Engineering, Energy and Architecture Set – Volume 9)

2021

BARDIOT Clarisse
Performing Arts and Digital Humanities: From Traces to Data
(Traces Set – Volume 5)

BENSRHAIR Abdelaziz, BAPIN Thierry
From AI to Autonomous and Connected Vehicles: Advanced Driver-Assistance Systems (ADAS)
(Digital Sciences Set – Volume 2)

BÉRANGER Jérôme
Societal Responsibility of Artificial Intelligence: Towards an Ethical and Eco-responsible AI
(Technological Prospects and Social Applications Set – Volume 4)

CORDELIER Benoit, GALIBERT Oliver
Digital Health Communications
(Technological Prospective and Social Applications Set – Volume 5)

DOUAY Nicolas, MINJA Michael
Urban Planning for Transitions

GALINON-MÉLÉNEC Béatrice
The Trace Odyssey 1: A Journey Beyond Appearances
(Traces Set – Volume 4)

HENRY Antoine
Platform and Collective Intelligence: Digital Ecosystem of Organizations

KOLOKOTSA Denia, KAMPELIS Nikos
Smart Buildings, Smart Communities and Demand Response
(Engineering, Energy and Architecture Set – Volume 8)

LE LAY Stéphane, SAVIGNAC Emmanuelle, LÉNEL Pierre, FRANCES Jean
The Gamification of Society
(Research, Innovative Theories and Methods in SSH Set – Volume 2)

RADI Bouchaïb, EL HAMI Abdelkhalak
Optimizations and Programming: Linear, Non-linear, Dynamic, Stochastic and Applications with Matlab
(Digital Sciences Set – Volume 1)

2020

ALAKTIF Jamila, CALLENS Stéphane
Migration and Climate Change: From the Emergence of Human Cultures to Contemporary Management in Organizations

BARNOUIN Jacques
The World's Construction Mechanism: Trajectories, Imbalances and the Future of Societies
(Interdisciplinarity Between Biological Sciences and Social Sciences Set – Volume 4)

ÇAĞLAR Nur, CURULLI Irene G., SIPAHIOĞLU Işıl Ruhi, MAVROMATIDIS Lazaros E.
Thresholds in Architectural Education
(Engineering, Energy and Architecture Set – Volume 7)

DUBOIS Michel J.F.
Humans in the Making: In the Beginning was Technique
(Social Interdisciplinarity Set – Volume 4)

ETCHEVERRIA Olivier
The Restaurant, A Geographical Approach: From Invention to Gourmet Tourist Destinations
(Tourism and Mobility Systems Set – Volume 3)

GREFE Gwenaëlle, PEYRAT-GUILLARD Dominique
Shapes of Tourism Employment: HRM in the Worlds of Hotels and Air Transport
(Tourism and Mobility Systems Set – Volume 4)

JEANNERET Yves
The Trace Factory
(Traces Set – Volume 3)

KATSAFADOS Petros, MAVROMATIDIS Elias, SPYROU Christos
Numerical Weather Prediction and Data Assimilation
(Engineering, Energy and Architecture Set – Volume 6)

MARTI Caroline
Cultural Mediations of Brands: Unadvertization and Quest for Authority
(Communication Approaches to Commercial Mediation Set – Volume 1)

MAVROMATIDIS Lazaros
Climatic Heterotopias as Spaces of Inclusion: Sew Up the Urban Fabric
(Research in Architectural Education Set – Volume 1)

MOURATIDOU Eleni
Re-presentation Policies of the Fashion Industry: Discourse, Apparatus and Power
(Communication Approaches to Commercial Mediation Set – Volume 2)

SCHMITT Daniel, THÉBAULT Marine, BURCZYKOWSKI Ludovic
Image Beyond the Screen: Projection Mapping

VIOLIER Philippe, with the collaboration of TAUNAY Benjamin
The Tourist Places of the World
(Tourism and Mobility Systems Set – Volume 2)

2019

BRIANÇON Muriel
The Meaning of Otherness in Education: Stakes, Forms, Process, Thoughts and Transfers
(Education Set – Volume 3)

DESCHAMPS Jacqueline
Mediation: A Concept for Information and Communication Sciences
(Concepts to Conceive 21st Century Society Set – Volume 1)

DOUSSET Laurent, PARK Sejin, GUILLE-ESCURET Georges
Kinship, Ecology and History: Renewal of Conjectures
(Interdisciplinarity Between Biological Sciences and Social Sciences Set – Volume 3)

DUPONT Olivier
Power: A Concept for Information and Communication Sciences
(Concepts to Conceive 21st Century Society Set – Volume 2)

FERRARATO Coline
Prospective Philosophy of Software: A Simondonian Study
(Technological Prospects and Social Applications Set – Volume 3)

GUAAYBESS Tourya
The Media in Arab Countries: From Development Theories to Cooperation Policies

HAGÈGE Hélène
Education for Responsibility
(Education Set – Volume 4)

LARDELLIER Pascal
The Ritual Institution of Society
(Traces Set – Volume 2)

LARROCHE Valérie
The Dispositif: A Concept for Information and Communication Sciences
(Concepts to Conceive 21st Century Society Set – Volume 3)

LATERRASSE Jean
Transport and Town Planning: The City in Search of Sustainable Development

LELEU-MERVIEL Sylvie, SCHMITT Daniel, USEILLE Philippe
From UXD to LivXD: Living eXperience Design

LENOIR Virgil Cristian
Ethically Structured Processes
(Innovation and Responsibility Set – Volume 4)

LOPEZ Fanny, PELLEGRINO Margot, COUTARD Olivier
Local Energy Autonomy: Spaces, Scales, Politics
(Urban Engineering Set – Volume 1)

METZGER Jean-Paul
Discourse: A Concept for Information and Communication Sciences
(Concepts to Conceive 21st Century Society Set – Volume 4)

MICHA Irini, VAIOU Dina
Alternative Takes to the City
(Engineering, Energy and Architecture Set – Volume 5)

PÉLISSIER Chrysta
Learner Support in Online Learning Environments

PIETTE Albert
Theoretical Anthropology or How to Observe a Human Being
(Research, Innovative Theories and Methods in SSH Set – Volume 1)

PIRIOU Jérôme
The Tourist Region: A Co-Construction of Tourism Stakeholders
(Tourism and Mobility Systems Set – Volume 1)

PUMAIN Denise
Geographical Modeling: Cities and Territories
(Modeling Methodologies in Social Sciences Set – Volume 2)

WALDECK Roger
Methods and Interdisciplinarity
(Modeling Methodologies in Social Sciences Set – Volume 1)

2018

BARTHES Angela, CHAMPOLLION Pierre, ALPE Yves
Evolutions of the Complex Relationship Between Education and Territories
(Education Set – Volume 1)

BÉRANGER Jérôme
The Algorithmic Code of Ethics: Ethics at the Bedside of the Digital Revolution
(Technological Prospects and Social Applications Set – Volume 2)

DUGUÉ Bernard
Time, Emergences and Communications
(Engineering, Energy and Architecture Set – Volume 4)

GEORGANTOPOULOU Christina G., GEORGANTOPOULOS George A.
Fluid Mechanics in Channel, Pipe and Aerodynamic Design Geometries 1
(Engineering, Energy and Architecture Set – Volume 2)
Fluid Mechanics in Channel, Pipe and Aerodynamic Design Geometries 2
(Engineering, Energy and Architecture Set – Volume 3)

GUILLE-ESCURET Georges
Social Structures and Natural Systems: Is a Scientific Assemblage Workable?
(Social Interdisciplinarity Set – Volume 2)

LARINI Michel, BARTHES Angela
Quantitative and Statistical Data in Education: From Data Collection to Data Processing
(Education Set – Volume 2)

LELEU-MERVIEL Sylvie
Informational Tracking
(Traces Set – Volume 1)

SALGUES Bruno
Society 5.0: Industry of the Future, Technologies, Methods and Tools
(Technological Prospects and Social Applications Set – Volume 1)

TRESTINI Marc
Modeling of Next Generation Digital Learning Environments: Complex Systems Theory

2017

ANICHINI Giulia, CARRARO Flavia, GESLIN Philippe,
GUILLE-ESCURET Georges
Technicity vs Scientificity: Complementarities and Rivalries (Interdisciplinarity Between Biological Sciences and Social Sciences Set – Volume 2)

DUGUÉ Bernard
Information and the World Stage: From Philosophy to Science, the World of Forms and Communications (Engineering, Energy and Architecture Set – Volume 1)

GESLIN Philippe
Inside Anthropotechnology: User and Culture Centered Experience (Social Interdisciplinarity Set – Volume 1)

GORIA Stéphane
Methods and Tools for Creative Competitive Intelligence

KEMBELLEC Gérald, BROUDOUX Evelyne
Reading and Writing Knowledge in Scientific Communities: Digital Humanities and Knowledge Construction

MAESSCHALCK Marc
Reflexive Governance for Research and Innovative Knowledge (Responsible Research and Innovation Set - Volume 6)

PARK Sejin, GUILLE-ESCURET Georges
Sociobiology vs Socioecology: Consequences of an Unraveling Debate (Interdisciplinarity Between Biological Sciences and Social Sciences Set – Volume 1)

PELLÉ Sophie
Business, Innovation and Responsibility (Responsible Research and Innovation Set – Volume 7)

2016

ANDRÉ Michel, SAMARAS Zissis
Energy and Environment
(Research for Innovative Transports Set – Volume 1)

BLANQUART Corinne, CLAUSEN Uwe, JACOB Bernard
Towards Innovative Freight and Logistics
(Research for Innovative Transports Set – Volume 2)

COHEN Simon, YANNIS George
Traffic Management
(Research for Innovative Transports Set – Volume 3)

EL FALLAH SEGHROUCHNI Amal, ISHIKAWA Fuyuki, HÉRAULT Laurent, TOKUDA Hideyuki
Enablers for Smart Cities

GIANNI Robert
Responsibility and Freedom: The Ethical Realm of RRI
(Responsible Research and Innovation Set – Volume 2)

GRUNWALD Armin
The Hermeneutic Side of Responsible Research and Innovation
(Responsible Research and Innovation Set – Volume 5)

LENOIR Virgil Cristian
Ethical Efficiency: Responsibility and Contingency
(Responsible Research and Innovation Set – Volume 1)

PELLÉ Sophie, REBER Bernard
From Ethical Review to Responsible Research and Innovation
(Responsible Research and Innovation Set – Volume 3)

REBER Bernard
Precautionary Principle, Pluralism and Deliberation: Science and Ethics
(Responsible Research and Innovation Set – Volume 4)

TORRENTI Jean-Michel, LA TORRE Francesca
Materials and Infrastructures 1
(Research for Innovative Transports Set – Volume 5A
Materials and Infrastructures 2
(Research for Innovative Transports Set – Volume 5B)

VENTRE Daniel
Information Warfare – 2nd edition

YANNIS George, COHEN Simon
Traffic Safety
(Research for Innovative Transports Set – Volume 4)

2015

BRONNER Gérald
Belief and Misbelief Asymmetry on the Internet

2009

VENTRE Daniel
Information Warfare